Radar for Meteorologists

or

You, too, can be a Radar Meteorologist, Part III

Thi

Ronald E. Rinehart
University of North Dakota

Rinehart Publications
P.O. Box 6124
Grand Forks, ND 58206-6124
United States of America

To Linda

Back cover photograph: Freezing rain storm detected by the UND C-band weather radar while operating in Kansas City, Missouri, on 14 February 1990. The left side shows Doppler radial velocity while the right side shows radar reflectivity factor. Elevation angle = 6.5°.

Color Figures 19-21 are based on NEXRAD images provided courtesy of INTELLICAST (http://www intellicast.com/) INTELLICAST is a registered trade-mark of WSI Corporation.

Cover designed by John Holland, Knight Printing
Printed by Knight Printing Company, Fargo, ND
First printing: September 1997
Second printing: July 1999

Available from:
Rinehart Publications
P.O. Box 6124
Grand Forks, ND 58206-6124
United States of America
701-772-7647
fax: 701-795-1914
email: radarwx@aol.com
homepage: http://www.aero.und.edu/~rinehart

ii

Table of Contents

Preface

Since the "good old days" when radar was first used to detect storms until now, radar has changed from a research tool used only by a few specialists to a tool used by many people on a daily basis. This transformation from an exotic and expensive research tool into a basic source of information for the masses has occurred for several reasons. One is simply the relative proliferation of radars across the country. Another is in the ability to communicate the data from place to place. Still another is that the information provided by radar is simply so useful that it is foolish for many people not to use it.

As the field of radar meteorology changed over the years, so, too, has the typical radar meteorologist. Based on photographs from some of the early radar conferences and actual observations of real, live radar meteorologists presenting papers at a number of radar conferences, I have watched the changes that have taken place over the years. Figure 1, first presented at the 22nd Conference on Radar Meteorology at Zurich, Switzerland (Rinehart, 1984), shows what the well-dressed radar meteorologist looked like over the years. Based on a quantitative analysis (least squares linear regression) on the amount of clothing and facial hair as a function of time, I was able to forecast what the well-dressed radar meteorologist of 2006 AD would look like. This forecast is shown in Fig. 2.

Preface

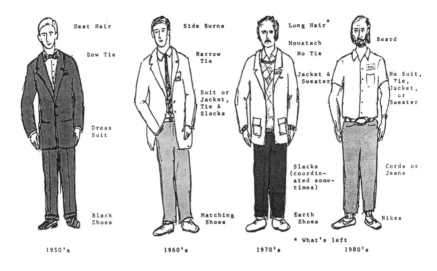

1950's   1960's   1970's   1980's

Figure 1   What the well-dressed radar
meteorologist wore in various years.

Figure 2   Forecast of what a radar
meteorologist in the year 2006 AD will look like.

Preface

There is a vast and rich number of publications in the field of radar meteorology. Much of this literature results from specific research studies into the use of radar and its application to various meteorological problems. Indeed, the majority of the publications related to radar meteorology are the results of research performed by hundreds of radar meteorologists over the past 40 years.

Much of the weather radar research results is found in the preprint volumes of the 26 conferences on radar meteorology that have been held since 1947; however, only the last 25 of these had conference proceedings. From time to time there have been additional summaries of the current state of the art. These include an early article in the *Compendium of Meteorology* (1952) and the recently published volume *Radar in Meteorology* (Atlas, 1990), both of which were published by the American Meteorological Society. Another would be the *Advances in Radar Meteorology* by Atlas in 1964.

In addition to the refereed literature, the Radar Conference Preprint volumes, and the special summary volumes, there have been a number of textbooks published by scientists and engineers for scientists and engineers. These include (roughly chronologically) Battan's 1959 *Radar Meteorology* and his 1973 revision *Radar Observation of the Atmosphere*; Skolnik's 1970 *Radar Handbook* and his 1980 *Introduction to Radar Systems*; Gossard and Strauch's 1983 *Radar Observation of the Clear Air*; and Doviak and Zrnic's 1993 *Doppler Radar and Weather Observations*. Obviously, there is a lot of information available in the field of radar meteorology.

So, why should there be another text on the subject? Much of the textual material now available is aimed at the research community with little practical information aimed at the real users -- front line meteorologists. My purpose in writing this text is to try to fill the gap between the instruction manual of a radar system and the theoretical treatment given in the most recently available texts. There is a need for an introductory text that not only gives the basic theory behind the wonders of radar but also gives some practical aspects in the use of radar to help the user understand what he is seeing and avoid some of the numerous pitfalls that await the unsuspecting user.

And, why is this "Part III"? My first attempt to summarize some practical radar techniques was aimed at an audience of one: my replacement on the Kericho, Kenya, hail suppression project in 1970. It included information on the quantitative use of radar, suggestions on how to recognize artifacts in the data (sidelobe tops), the use of ground targets for radar system checks, and a number of other site-specific and project-specific problems. While I am hopeful that it was helpful, I really don't know that for a fact.

Part II of "You, too, can be a radar meteorologist" was a several page discussion of how to do dual-Doppler processing of data from the Lincoln Laboratory FL2 radar and the University of North Dakota (UND) radar. The audience of this version was approximately a dozen "users" at Lincoln Lab in 1986. Again, not all the users used it, but some might have found it helpful.

And now part III. This one is intended to be more complete and provide a broad (but perhaps shallow) coverage of the current state of knowledge of *practical* radar meteorology. I emphasize "practical" because that seems to be missing in many available sources now days. It is not difficult to find all sorts of manipulations, derivations and aggravations awaiting those reading the research-oriented literature. But there is little information related to such mundane things as how to detect the presence of sidelobe tops, second-trip echoes, or the uncertainties associated with radar measurements. I hope to fill this gap and provide a useful reference for the day-to-day users of radar data.

A second motivation for writing this is for use as a text in an undergraduate course in radar meteorology. Graduate students are usually well prepared to cover the gory details of Fourier analysis, Bessel functions, and similar mathematical manipulations. Undergraduates generally don't need or want that kind of detail. They need a more fundamental level of information. It still needs to be physically based and well grounded theoretically, but it can and should avoid some of the derivations and detail found elsewhere. I hope this will prove to be that kind of text.

Finally, with the installation of WSR-88D (formerly called NEXRAD, NEXt generation RADar) systems across the country, there will be a lot of new users and uses for radar data. What are the advantages of NEXRAD data? What are the limitations of NEXRAD data? How can it best be used? NEXRAD is going to make more and better radar data an everyday occurrence. We - users, teachers, and the general public - will all benefit from its use. But we will all have some learning to do before we make the best possible use of it.

In appreciation, I would like to thank the following individuals: For helpful reviews of portions of the text: Mike Poellot, Tony Grainger, Don Burrows, Erwin Prater, and Dave Bernhardt, all from UND. For generously supplying data and doing processing over the years: Al Borho (UND), Scott Kroeber (UND), Mark Isaminger, Bob Hallowell (both from Lincoln Lab), and Don Meissner (UND). Thank you, gentlemen.

Preface

So, with all of this out of the way, let us begin our study of radar meteorology. And -- enjoy!

July 1990   Ronald E. Rinehart

Preface to the Second Edition

In many ways, this second edition is really the first edition to the general public. The first edition was distributed to a very limited number of people, including students here at the University of North Dakota. In other ways, however, this is truly a second edition. The most significant change between the first and second editions is the addition of color figures of radar data. Al Borho and Paul Kucera's help in assembling and photographing these is greatly appreciated.

Also, thanks are due to Ron Alberty and Rudy Schaar of the NEXRAD office in Norman, Oklahoma, for supplying the color figures of WSR-88D radar data and the photograph of the OSF radar site at Norman, Oklahoma. I would also like to thank Bob Hallowell, MIT Lincoln Laboratory, who gave me some very useful comments on the entire text.

Aside from the switch to color, most of the other changes were additions to the text, some rearrangement of material into a more logical order, and the removal of "all known errors." Of course, new ones have likely taken their place. Since this is as close to desktop publishing as a book likely gets, I have to take sole responsibility for all of these.

It is my sincere hope that this book will prove useful for those interested in the practical aspects of radar meteorology. Again, enjoy!

April 1991   Ronald E. Rinehart

Preface to the Third Edition

The Third Edition of *Radar for Meteorologists* is a more or less complete revision of the Second Edition. Some material was deleted but more was added.

During the six years since the Second Edition came out, virtually the entire network of NEXRAD radars has been installed and come on line. It is now possible to get reflectivity data from all WSR-88D radars in the conterminous United States over the Internet with as little as ten minutes or so delay. This makes it possible to watch evolving weather situations from anywhere in almost real-time. Additional examples of NEXRAD data are included in the color figures.

Again, I would like to thank some individuals for their contributions to this revised text. My sincere heartfelt thanks especially go to Andy Detwiler and Rodger Brown who graciously agreed to review major parts of the text. Their help went far beyond what mere volunteers should ever be expected to do. And Daran Rife also reviewed a significant portion of the text. The Flood of '97 in Grand Forks threw a wet blanket on my plans for completing this in a timely fashion. These gentlemen helped improve the text while inspiring me to keep at it through completion. Thank you!

Mr. Jim Belles, Grand Forks National Weather Service Forecast Office, kindly provided the VAD image used as Fig. 11.2 and the reflectivity and velocity images used for Color Figures 17 and 18. Thanks.

I also want to thank Bob Clymer, a graduate student at the University of Alabama at Huntsville, who alerted me to the bird echoes shown in Color Figures 20 and 21. I think they are a great source of amusement and information; I have thoroughly enjoyed watching them over the past couple of years.

Finally, this is again the product of desk-top publishing, so _ALL_ errors herein are my responsibility. Feel free to point out errors or offer suggestions for future changes. Your contributions can keep this a useful text for use by meteorologists, researchers, technicians and anyone else who is interested in a basic understanding of how weather radar operates.

August 1997   Ronald E. Rinehart

Chapter 1

Introduction

History of radar

The history of radar is very closely linked to the history of radio. The very word "radar" suggests its origin in radio: "Radio Detection And Ranging." This history includes the various practical and theoretical discoveries of the 18th, 19th and early 20th centuries that paved the way for the use of radio as a means of communication. Robert Buderi (1996) recently wrote an excellent book that details the history of radar. *The Invention That Changed the World* is a fascinating account of the history of radar, including the people who did it. Anyone who is interested in learning about the development of radar and how it has "changed the world" will find it extremely interesting.

Radar has its roots in radio. Once radio was well established as a communications tool, there were a number of events that built upon that experience for radar. Buderi gives the account of one of the events that helped pave the way for radar as follows:

> "...In September 1922, the wooden steamer *Dorchester* plied up the Potomac River and passed between a transmitter and receiver being used for

experimental U.S. Navy high-frequency radio communications. The two researchers conducting the tests, Albert Hoyt Taylor and Leo C. Young, had sailed on ships and knew of the difficulty in guarding against enemy vessels seeking to penetrate harbors and fleet formations under darkness. Quickly putting the serendipitous finding together, the men proposed using radio waves like a burglar alarm, stringing up an electromagnetic curtain across harbor entrances and between ships. But receiving no response to the suggestion, and with many demands on their time, the investigators let the idea wither on the vine."

Interestingly, Young and Taylor had a similar experience in 1934 when and aircraft interrupted their radio signals. As a result of that, they proposed the use of using pulses of energy for target detection[1].

The development of radar-like instruments was done simultaneously in a number of countries, including the United States, Great Britain, France, Italy, Germany, Holland, Russia and Japan. Early in World War II the United States and Great Britain joined forces to share the development of radar.

One of the key inventions that made radar successful was that of the magnetron transmitter tube by John Randall and Henry Boot at the University of Birmingham. The magnetron is the tube which made it possible to have moderately light-weight radars which operate at the higher, microwave frequencies where

---

[1] In earlier editions of this book, I related a version of this story. What makes it especially interesting to me is that Dr. Taylor served as the chairman of the Physics Department at the University of North Dakota before commencing his Naval research activities.

radar has some advantage over lower frequency, longer wavelength systems. The use of high frequency signals made it possible to reduce the size of the antenna considerably. This enabled radar to determine the direction to a target much more accurately. Since the targets of interest during WWII were primarily aircraft, the ability to find them accurately was a very strong motivation for improving this capability.

By the end of WWII, radar had been thoroughly developed and utilized very successfully, especially by the allied forces. After the war, much of the surplus military equipment became available for civilian use. Those interested in studying the weather with radar were some of the first to acquire surplus radars and put them to research uses.

Since World War II, radar has undergone an evolutionary development. Improvements have been made in all components of radar, from the transmitters to the antennas to the receivers, displays and processors. Inventions in other fields (such as the transistor) soon found their way into radar systems. New types of antenna systems such as phased arrays were developed by the military and are beginning to find civilian uses as well. And, of course, the application of computers to radar has shifted many of the new developments in radar from the hardware to the software. It is now possible to provide automatic warnings by processing radar data by computers.

One of the greatest advances in radar technology, at least as far as weather uses are concerned, was the development of Doppler techniques. Doppler radars not only detect and measure the power received from a target, they also measure the speed of the target toward or away from the radar, i.e., the *radial* velocity of the target. If more than one target is present in the radar's sampling volume, the velocity of each individual target will be received. In the case of detecting storms with radar, there may be billions of raindrops in a single sampling volume. In this case, a Doppler radar will not give a billion different velocities, but it can determine the relative number of targets with each velocity. This can provide useful information on the distribution of drop sizes present in the sample volume.

Another advance since the war has been the utilization of polarization information. As we will see later, the ability of some radars to detect information from the polarization of the received signal can provide information about the kind of particles being detected (e.g., ice or water), their shapes, and sizes.

How does radar work?

In Chapter 2 we will learn more about the workings of radar, but it might be useful to have a quick overview of the workings of a radar so the following discussion of radar types will make more sense. For this introductory description, let's consider a typical weather radar since they are what this text is primarily about. Other radars may operate differently, but all have certain characteristics in common.

Radar basically consists of four main components. These include a transmitter to generate the high-frequency signal, an antenna to send the signal out into space and to receive the echo back from the target, a receiver to detect and amplify the signal so it is strong enough to be useful, and some kind of display system to allow people to see what the radar has detected.

Early radars often used a transmitter that generated a signal continuously. Because of that, the transmitter was connected to its own antenna. The receiver had to be connected to another antenna. And often, the transmitting and receiving antennas were separated by some distance so the transmitted signal would not overpower the receiver. Figure 1.1 illustrates a bistatic radar detecting the ionosphere.

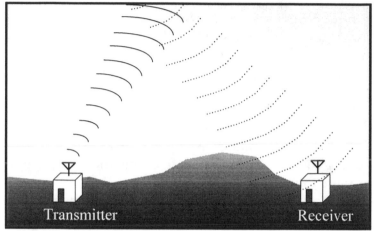

Figure 1.1 *Bistatic radar sending its signal from one location and receiving it at another location. In this case, the target could be the ionosphere.*

Modern weather radars use a single antenna. Weather radars send out short pulses of energy and then wait awhile so the signal can travel out at the speed of light, hit a target, and return back to the antenna. After an appropriate wait, another pulse is sent out. And another and another, etc. The energy travels out at the speed of light, so it does not take very long for the signal to travel several hundred miles.

The antenna of the radar usually rotates about a vertical axis, scanning the horizon in all directions around the radar site. Even though the antenna may be scanning at a speed of from 10°/s to as much as 70°/s (up to 10 RPM or faster), the speed of light is _much_ faster. It only takes 2 ms (milliseconds) for a radar signal to travel out 300 km and back. If the antenna is rotating at 10 RPM, the antenna will move less than a tenth of a degree during this time. So, from the human perspective, the signal went out and back with the antenna virtually stationary. The radar repeats its transmission/listening cycle several hundred to a couple of thousand times each second. So, we mere mortals watch the antenna moving smoothly and fairly rapidly, but we often don't appreciate the fact that the signal is going out and back very long distances at speeds that are almost incomprehensible to the human mind. We see the radar painting echoes smoothly as the trace sweeps along on the radar display, and it looks like it is not doing much. In reality, it is doing thousands of times more things than we can see happen.

Meteorological radars can also aim their antennas above the horizon. This is a useful thing to do if we want to determine how high a storm is or to detect echoes that are above the surface. Some radars operate in what is called a volume-scan mode. The antenna will do a full circle at one elevation angle, then tilt up a degree or two and do another circle. The antenna will complete as many as 10 to 20 different elevation angles and then repeat the whole cycle in a period of perhaps four to six minutes. By doing this, it is able to scan the entire volume surrounding the radar, collecting very useful data on the vertical structure and distribution of storm mass.

Some radars operate by scanning up and down while slowly rotating in azimuth. These height-finding radars provide a different view of storms and are a useful complement to conventional radars which scan mostly horizontally.

Finally, once the signal has been detected and amplified by the receiver, the signal goes to the display device so it can be presented in such a way that the human brain can see and understand what the radar has detected. Most weather radars display their data on a map-like display that also shows such things as county and state boundaries, the locations of cities and other important features. The radar data are usually color coded so we can determine which storms are strong and dangerous and which are more benign and unimportant.

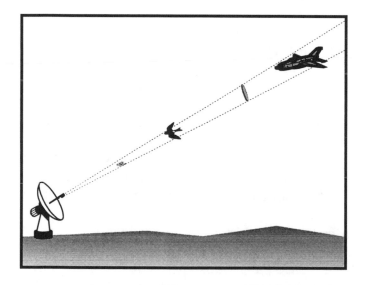

Figure 1.2 A radar looking in the direction of an insect, a bird, and an aircraft. Also shown is a short pulse of energy traveling outward at the speed of light (look fast). Note that nothing in this drawing is to the same scale!

Now, with that as background, let's consider some of the types of radars that exist today.

Chapter 1

Types of radar

Almost anything that can be classified can be subclassified. When we are only vaguely familiar with a subject, we will know only the most basic membership categories. For example, when we think of giraffes or zebras or other animals, we think there is only one kind of each. Actually, there are three kinds of giraffes (Masai, Reticulated, and Rothschild's) and two kinds of zebras (common and Grevy's). So it is with radars; there are many kinds.

A quick list of radars could include the following: Monostatic and bistatic radar, CW and pulsed radar, Doppler radar, FM-CW radar, wind profilers, approach control and aircraft surveillance radars, airborne radar, shipboard radar, weather radar, dual-wavelength radar, polarization-diversity radar, etc. Each of these is designed for a specific purpose and usually does its particular job very well; on the other hand, it may not be useful at all for some other purposes. Let's consider some of these and how they operate and differ from others on the list.

Monostatic vs. bistatic radar: When we think of weather radar, we are usually thinking about what could be called a monostatic, pulse radar. The terminology "monostatic" and "bistatic" has to do with where the radar's receiving antenna is in relationship to its transmitting antenna. Most radar systems use a single antenna for both transmitting and receiving; the received signal must come back to the same place it left in order to be received. This kind of radar is a monostatic radar.

A bistatic radar, on the other hand, has two antennas. Sometimes these are side by side, but sometimes the transmitter and its antenna at one location and the receiver and its antenna are at another (see Fig. 1.1). Of course, both antennas must be aimed at the same volume in space for the signal to make it from one to the other, but the distance between them can be quite far. In this kind of radar the transmitting radar system aims at a particular place in the sky where a cloud or other target is located. The signal from this point is scattered or reradiated in many directions, much of it being in a generally forward direction. The receiving antenna and the remainder of the system must be aimed toward the same region so it can detect the signal. The military operate "over-the-horizon" radars that have widely spaced transmitters and receivers. These OTH radars can detect aircraft and missiles at distances of several thousand kilometers.

Continuous Wave vs. pulsed radar: The second qualifier used in the description of a conventional weather radar was "pulse". This means that the radar transmits a short pulse of electromagnetic radiation and then waits for an echo from a target. An alternative to this kind of system is to transmit continuously (CW means Continuous Wave transmission). When this kind of radar is used, the received signal coming back to the receiver is also continuous. The only way the radar can detect a target is for the received signal to be different from the transmitted signal in some way. One way they can be different is if the target is moving. Then there is a slight shift in the frequency of the reradiated signal which is proportional to the speed of the target relative to the radar. CW radars are used, for example, by police officers to measure the speed of traffic.

Doppler radar: One of the most important things that radars can measure was mentioned in the above paragraph: the velocity of a target. The principle by which these radars operate was first discovered by the person after whom these radars are named, Christian J.

Chapter 1

Doppler (1853). Doppler discovered that the shift in frequency caused by moving sources of sound was directly proportional to the speed of the source. The classic example of this is when a train approaches a crossing where a stationary observer listens to its whistle. The frequency of the sound will shift from one frequency to a lower frequency as the train passes. Exactly the same principle applies to electromagnetic radiation from a radar. In this case the radar is stationary but the target is moving.[2] If the target moves toward the radar, the frequency is increased; if it is moving away, the frequency is reduced. A Doppler radar compares the received signal with the frequency of the transmitted signal and measures the frequency shift, giving the speed of the target quite easily. The Doppler principle is applied in several different kinds of radars, including weather radars, police speed radars, and wind profilers.

Wind profilers are a relatively new kind of meteorological radar which operate at moderately low frequencies. Because of the wavelengths chosen, these radars are capable of detecting the wind between the surface of the earth and heights up into the lower stratosphere on an almost continuous basis. The potential for improved weather forecasts in the future is very great because of these new wind profilers. Currently the National Weather Service sends up weather balloons at a hundred or more sites twice a day every day. Wind profilers, however, can give the same

---

[2] Occasionally problems arise because the radar antenna must move during normal scanning. If conditions are right, this can give rise to echoes detected by the radar's sidelobe pattern that appear to have velocity even though the actual target may not be moving at all.

every day. Wind profilers, however, can give the same kind of wind information about every 5 min, 24 h a day. This will make it possible to detect much better information about winds in the troposphere and lower stratosphere where most of our weather occurs.

Radars used in Aviation:

ARSR: The Federal Aviation Administration (FAA) operates several different kinds of radars to assist the aviation industry. One is to detect en route aircraft flying the nation's airways. These radars operate at L-band (20-cm wavelength; see Table 3.1, Ch. 3) and are designed to detect aircraft at long ranges from the radar at fairly rapid intervals. One of the improvements of the ARSR-4 over the older ARSR-3 radars is that the new radars give a three-level weather map in addition to the normal aircraft position information. This makes the ARSR-4 a possible additional source of weather radar data for various uses. Appendix E gives the specifications of the ARSR-4 radar.

ASR: A second kind of FAA radar is the airport surveillance radar (ASR) type approach-control radar. The ASR-8 radars are currently being used at most major airports in the country (see Appendix E). Over the past decade or so, however, the FAA has been supporting the development of a newer version, the ASR-9. The ASR-9's are functionally similar to the older ASR-7's and -8's but also include the ability to detect the Doppler velocity of weather targets. These radars are useful in providing more detailed information on the positions of aircraft as they approach an airport to land. The ASR-9 radars have a modular data processing channel for automatic detection of windshear, thunderstorm microbursts, and gust fronts. This capability provides windshear warnings at airports radar systems. Deployment of the new ASR-9 radars should be completed by 2005 AD. The specifications of the ASR-9 are given in Appendix D along with those of some other meteorological radars.

TDWR: A third type of radar for use in the aviation community is the Terminal Doppler Weather Radar (TDWR). TDWR's are a new terminal Doppler weather radar which will detect microbursts, gustfronts, wind shifts, and precipitation. This radar will be used to alert aircraft in the terminal area of hazardous weather conditions and to provide advanced notice of changing wind conditions to permit timely changes of active runways. Over the past quarter century, more than 500 people have been killed because of microbursts. Microbursts can occur with or without precipitation and are particularly hazardous to aircraft which are taking off or landing. TDWR's will be located near the airport and have their scanning modes optimized for microburst and windshear detection. By the end of the century, 47 TDWR systems will be deployed across the country. See Appendix D for specifications of the TDWR.

ASDE: Radars are also used at some airports to follow aircraft on the ground. The airport surface detection radar (ASDE-3) detects aircraft on taxiways, runways, and elsewhere. Controllers can use data from these radars to turn lights on and off to direct aircraft from one taxiway to another and prevent mishaps on the ground.

Radar transponders: Many aircraft (all commercial aircraft) carry transponders which can tell when a radar is aimed in their direction. When a radar signal is detected, the transponder transmits a signal which usually includes the altitude of the aircraft (as determined by an on-board altimeter). The receiver at the radar site receives this altitude-encoded signal and then displays the aircraft position as determined by the radar (i.e., its position based on distance and azimuth from the radar) along with the altitude and identification transmitted by the aircraft. This helps the FAA maintain both horizontal and altitude separation between the numerous aircraft in the sky at any one time.

On-board radars: Weather radars used on board small, general aviation aircraft must be designed to fit within limited space and weight constraints. The most severe restriction is on the size of the antenna that can be carried inside the radome. As will be seen later, the size of the antenna is related to the frequency of the radar signal. Short wavelength radars can use smaller antennas than longer wavelength radars for the same angular resolution. However, by going to a short wavelength to get better angular resolution, other problems are introduced. The biggest problem is that short wavelength radar signals are much more attenuated than those from longer wavelengths. Attenuation can cause the true strength of potentially dangerous storms to be grossly underestimated or missed completely, so attenuation is a serious problem for airborne radars. We will discuss attenuation in more depth later. As we will see, there are a number of constraints in the choice of some radar parameters. Practical (i.e., *real*) radars are always a compromise between several conflicting parameters.[3]

---

[3] Archie Trammell, AJT, Inc., gives a series of seminars on the use of radar onboard aircraft. In one of his presentations he related the story of the ultimate ideal, ultrasafe aircraft. In his story, the pilot and copilot are flying along. The copilot turns to the pilot and says, "Sir, we're having a problem with the

Chapter 1

Weather radars:

Last, but certainly not least, we have weather radar - the subject of this book. There are many different kinds of weather radars in use today. Weather radars come in a variety of shapes and sizes, from small X-band radars used at television stations, small airports, and other locations, to the 178 WSR-88D (NEXRAD) S-band radars used by the National Weather Service, the FAA and the Department of Defense. Many modern weather radars have Doppler capability. A few have the ability to change polarization. And some combine those capabilities with still more sophisticated capabilities. We will discuss weather radar and the various flavors they come in throughout this text, so we can conclude this introduction by simply saying that weather radars are one of the most important weather tools in use today.

---

number ten engine." The pilot asks the copilot, "Is that on the right side or left side?" Maybe we could add a 28-ft diameter S-band radar antenna to such an aircraft to make it even safer!

Chapter 2

Radar Hardware

A radar system (or, more simply, a radar) consists of several subsystems. Each of these subsystems itself consists of many components and even smaller subsystems. It is beyond the scope of this text to describe in detail all the inner workings of a radar. Manufacturers usually provide complete sets of schematic diagrams, operating manuals, theory of operation, and other references that can be consulted for all the nitty-gritty details on a specific radar. Even if you are not a radar engineer, it would be worth spending some time examining these manuals because they contain a wealth of information on the capabilities and limitations of the radar you may be using. If you are a user of radar data and have no direct access to the radar itself, you should still be aware that these manuals exist. Manufacturers put a lot of effort into writing these manuals. They are an excellent reference to learn more about how a specific radar operates.

For our purposes, however, we only need to consider the main components, those that are found on all weather radars. Figure 2.1 shows a very simple block diagram of a radar. Let's examine each of these components in a little more detail.

Chapter 2

Transmitter

The source of the electromagnetic radiation emitted by a radar is the transmitter. It generates the high frequency signal which leaves the radar's antenna and goes out into the atmosphere. There are several kinds of transmitters used in modern radars, each of which has its own advantages and disadvantages. The three most important kinds for meteorological radars are the magnetron, the klystron, and solid-state transmitters. Each of these can be designed to optimize particular characteristics.

The magnetron tube for generating microwaves was invented by John Randall and Henry Boot in November 1939 (Buderi, 1996). Randall and Boot were working on the development of radar for the British. Magnetrons proved to be one of the most important developments of radar for World War II. Their small size and high power output made them ideal for airborne use. Magnetron transmitters are now used at a variety of radar frequencies and can generate transmitted signals in excess of 500 kW.

Figure 2.2 shows a schematic image of a cavity magnetron. The cavity portion of the magnetron is sandwiched between a very strong magnet so that the magnetic field is oriented into the page (for the top view). During World War II, magnetrons used by the British contained six cavities while those built by the Americans used eight cavities. Magnetrons used today typically have more than six or eight cavities.

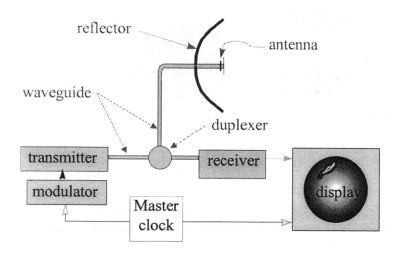

Figure 2.1 Block diagram of a simple radar.

Klystron transmitters have several advantages over magnetron transmitters. Klystrons, while they are usually bigger and bulkier, are true amplifiers. As such, it is easier to control the waveform of their transmitted pulses. Further, they are usually capable of transmitting more power than magnetrons (up to 2 MW or more) and their output signals are of purer frequencies. This latter characteristic is particularly suited to measuring the speed of motion of detected targets.

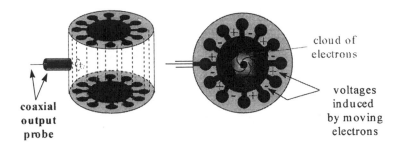

coaxial
output
probe

cloud of
electrons

voltages
induced
by moving
electrons

Figure 2.2. Cavity magnetron showing the central cathode (negative) and surrounding anode (positive). The probe takes the energy from the magnetron and radiates it into the waveguide. The electrons emitted by the cathode (negative) move outward toward the anode (positive). The magnetic field (perpendicular to the plane of the page), however, causes the electrons to spiral in their path. As the electrons pass the openings to the cavities, they create a kind of electronic "whistle", generating the microwave frequency of the magnetron. This signal is radiated into the waveguide by the probe at left.

Solid-state transmitters usually have much lower transmitter powers than most radar transmitters. A single solid-state transmitter might transmit as little as 50 W of power. However, by combining multiple solid-state transmitters into an array and controlling the timing of each element appropriately, it is possible to achieve useful power outputs. Solid-state transmitters have not been used much for ground-based meteorological radars, but at least one manufacturer is selling an aircraft radar using this approach.

Modulator

No matter what kind of transmitter is used in the radar, it is usually controlled by another electronic device called the modulator. The purpose of the modulator is to switch the transmitter on and off and to provide the correct waveform for the transmitted pulse. That is, the modulator tells the transmitter when to transmit and for what duration.

The modulator also serves another function. It stores up energy between transmitter pulses so when it is time for the transmitter to fire, it will have a storehouse of energy available for its use.

Master Clock/Computer

Since weather radars are built so they can operate in a variety of ways, there must be some kind of interface between the operator and the radar to translate our wishes into radar commands. The control panel of the radar provides for a number of choices and a means to select them so we can communicate with the radar.

Inside the radar there must be some circuits to convert our selections into control signals for each function selected. For example, the operator usually selects the range to display, the elevation angles of the antenna, the azimuths to scan, and often the number of pulses transmitted each second and how long each of these pulses will be.

In older radars, the device which did much of this was called the master clock. It would generate all of the appropriate signals and send them to the appropriate components of the radar. In modern radars, the function of the master clock has been taken over by the ubiquitous computer. Computers now control radars just as they control many other parts of modern technology.

Besides controlling the operation of the radar, computers (also called signal processors) control the processing of the received data. Signal processors can take the incoming signal, run data quality checks on the data, average the data, convert it into new products, store it for later replay, transmit it to remote locations, and display it for human consumption. Nowadays, the computer attached to a radar is probably more important than the hardware. In fact, radar hardware has not changed nearly as much in recent years as has the software used with the radar.

One of the main functions of the master clock/computer is to control how often and how long the transmitter transmits. The rate at which the radar transmits is called the pulse repetition rate or pulse repetition frequency (PRF). PRF is usually measured in pulses or cycles per second or in hertz (1 Hz = 1 cycle/second). PRF's can be as low as 200 Hz and as high as 3000 Hz for various radars. Older, non-Doppler, ground-based weather radars used to operate with PRF's on the order of 150 to 300 Hz. Doppler weather radars – those capable of detecting the speed of targets moving toward or away from the radar -- typically operate with PRF's on the order of 700 to 3000 Hz. Weather radars used on board aircraft typically operate at 500 to 1500 Hz.

The duration of the transmitted signal goes by either of two different names. If measured in units of time, we call it the pulse duration (τ); if measured in units of distance, we call it pulse length (h). Typical pulse durations are from 0.1 to 10 μs (1 μs = 10^{-6} s). We can easily convert pulse duration into pulse length using *distance = rate • time* where *distance* is the pulse length *h*, the *rate* is the speed of light *c*, and the *time* is the pulse duration τ.

Waveguide

Figure 2.1 shows that the conductor connecting the transmitter and the antenna is waveguide. Regular wires work fine for conducting electricity and low-frequency signals. When electricity was first discovered, wires were all that was needed. As radio developed and higher and higher frequencies came into use, people discovered that simple wires were very lossy, i.e., too much energy was lost to make regular wires useful. Radio engineers soon found that a better way to carry radio-frequency signals was through special conductors called coaxial cables. Coaxial cable contains a center conductor surrounded by insulation and then by a layer of shielding conductor (and finally another layer of insulation). Coaxial cable works well at many radio frequencies.

At microwave radar frequencies, however, even coaxial cable is too lossy. To avoid these losses, another kind of conductor was invented which is quite efficient at carrying radar signals. This conductor is called waveguide. It is usually a hollow, rectangular, metal conductor whose interior dimensions depend upon the wavelength of the signals being carried (see Fig. 2.3). Waveguide is put together much like the copper plumbing in a house. Long pieces of waveguide are connected together by special joints to connect the transmitter/receiver and the antenna. This allows the transmitter and receiver to be located at one place while the antenna is mounted elsewhere, usually up on a tower for conventional, ground-based weather radar.

Since there is seldom a straight line between the transmitter and the antenna, waveguide also has to be able to conduct its signals around corners. There are a number of special pieces of waveguide to account for this need. The signal conducted inside the waveguide consists of both an electric and a magnetic component (see the following chapter for a discussion of the characteristics of electromagnetic radiation). The cross section of waveguide is usually rectangular rather than square. The longer dimension is the direction of the electric field while the shorter direction is the direction of the magnetic field. Waveguide can bend in either of two directions: in the direction of the electric field (called an E-plane bend) or in the direction of the magnetic field (an H-plane bend). Given a choice, it is better to use E-plane bends rather than H-plane bends since the losses in H-plane bends are greater than those of E-plane bends.

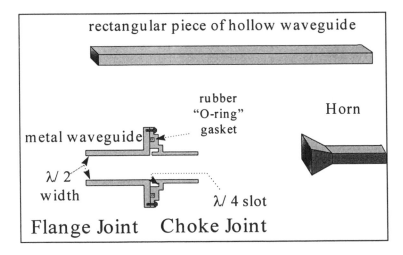

Figure 2.3. Waveguide (top), feedhorn (lower right), and a waveguide joint (lower left).

Another kind of waveguide is flexible waveguide. This consists of a waveguide that is sort of like a gooseneck lamp. The metal part of the waveguide can bend to accommodate slight misalignment in waveguide or to allow for slight movement between adjacent components. The outside of flexible waveguide is usually rubber coated to make it air and water tight.

A final special form of waveguide is a rotary joint. Most weather radars have the transmitter and/or receiver located on the ground while the antenna is located on a tower. Waveguide is used to connect the components. Antennas on weather radars must rotate so the antenna can scan horizontally (i.e., azimuthally) and in elevation. Rotary joints are used between the waveguide fixed to the radar tower and the waveguide fixed to the antenna. A second rotary joint is used so the antenna can scan up and down while the pedestal does not.

Waveguide comes in straight pieces which must be assembled into the final run of waveguide. The manufacturer usually puts all the waveguide together when the radar is installed at its final site. When the distance between two points is too far to reach with a single piece of waveguide or when a long piece must be cut for a short distance of waveguide, connectors must be attached to the waveguide to allow the pieces to be connected together. Figure 2.3 shows the ends that are attached together. Both a flange joint and a flat joint must be used to connect the waveguide properly. If two flat joints are connected, and even a very tiny crack exists, energy will be lost. To avoid that, the combination of a flat and a flange joint solve this problem. The flange joint contains two grooves. The groove near the outside is simply a grove for a rubber O-ring gasket to make the waveguide air tight. The inner groove is made a quarter of a wavelength long. This length causes energy entering the groove to be reflected back into the waveguide exactly in phase with the energy passing that point. The result is that the joint looks to the radar waves as if it is a perfect conductor and effectively prevents the loss of energy. If you ever assemble waveguide, be very careful that you do not connect two flange joints together. Doing that makes the equivalent of a half-wave groove. That is like a short circuit to radar waves! Waveguide joints are usually made with threads in the flat and a hole in the flange joint to make it impossible to inadvertently connect two flats or two flanges together.

Waveguide is an excellent conductor of microwave signals, but it is not perfect. Each waveguide component introduces losses. Skolnik (1980) discusses each of these losses in detail. Fig. 2.4 shows the losses of waveguide as a function of frequency. Notice that each waveguide size fits a range of frequencies. Also note that there is a limited number of waveguide sizes available. For a NEXRAD radars, for example, the antenna is typically 60 to 70 ft from the transmitter/receiver cabinet; the two-way loss for this length of waveguide would be just over 1 dB.

Antenna

Antennas are one of the most important components of a radar. At this point I feel a little like TV weatherman Willard Scott. Every time he broadcasts from a different city, he says that "this is my favorite city in the whole country." Now that we are ready to discuss antennas, I have to say that antennas are one of my favorite parts of a radar!

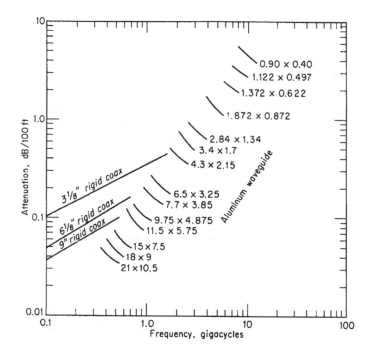

Figure 2.4 Waveguide attenuation as a function of waveguide size (dimensions are given in inches next to each particular size) and frequency. From Skolnik, 1980, Introduction to Radar Systems, with permission from McGraw-Hill. Inc.

The antenna is the device which sends the radar's signal into the atmosphere. Most antennas used with radars are *directional*; that is, they focus the energy into a particular direction and not in other directions. One of the great advantages of radar is its ability to determine the direction of a target from the radar. It is the ability of a radar's antenna to aim energy in one direction that makes it possible to locate targets in space.

An antenna that sends radiation equally in all directions is called an **isotropic antenna**. It can be compared to the light from a candle. A candle's light is approximately the same brightness in all directions, except, of course, directly below the candle. For weather radar, transmitting a signal equally in all directions would usually not be very useful. Instead, radars are more like flashlights. Flashlights put a shiny reflector behind the light bulb to direct the light in a specific direction.

Weather radars usually have both an antenna and a reflector (see Fig. 2.5). The *real* antenna is the radiating element which transmits the radar signal into the atmosphere toward the reflector that then reflects and directs the signal away from the radar. Most weather radars use a feedhorn as the true antenna although some use dipoles or other radiating elements.

The shape of the reflector determines the shape of the antenna beam pattern. Most meteorological radars have reflectors which are parabolic in cross-section and circular when viewed from the front or back; naturally, they are called circular parabolic reflectors. The beam pattern formed by a circular parabolic reflector is conical and usually quite narrow, typically 1° in width for the mainlobe of the pattern. The bigger the reflector, the better it is able to direct the signal and the narrower the beam of the antenna.

But there are other kinds of antennas used for meteorological radars. Figure 2.5 shows a Cassegrain-feed antenna. In this design, the actual antenna is located at the end of the tube coming out of the center of the main parabolic reflector. The signal is aimed at a hyperbolic-shaped subreflector. The signal then reflects back to the main parabolic reflector and then out into space. The received signal follows the reciprocal path

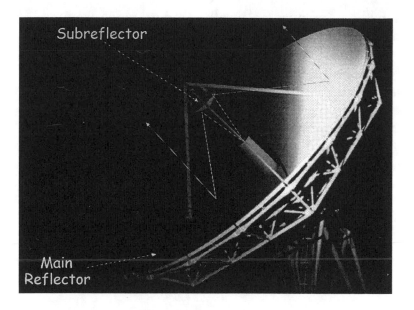

Figure 2.5   Photograph of a Cassegrain antenna with its main parabolic reflector, a hyperbolic subreflector.   The thin white lines indicate the paths a couple of representative rays might follow for a transmitted signal.

from space to the main reflector to the subreflector to the feedhorn to the receiver. One of the advantages of the Cassegrain feed is that they usually have better sidelobes for a given size antenna. The antenna beam pattern shown in Fig. 2.9 is from a Cassegrain antenna.

There are a number of things we need to know about an antenna. One is the wavelength it is designed for. The radar transmitter determines this parameter; the antenna must match the transmitter's wavelength.

A second parameter of interest is the size of the reflector. For circular parabolic reflectors, this is its diameter. Antennas on weather radars range from as small a foot to as much as 30 ft in diameter.

Another measure of importance to radar antennas is the gain of the antenna. The gain g of an antenna is the ratio of the power that is received at a specific point in space (on the center of the beam axis, i.e., at the point where the maximum power exists) with the radar reflector in place to the power that would be received at the same point from an isotropic antenna. This is a unitless ratio since it is one power divided by another power and units cancel. In equation form, gain is defined as

$$g = \frac{p_1}{p_2}$$

(2.1)

where p_1 is the power on the beam axis with the antenna and p_2 is the power from an isotropic antenna at the same point.

Usually antenna gain is measured logarithmically in decibels (see Appendix A for a more complete discussion of logarithmic units). A power ratio in decibels is defined as

$$P = 10 \log_{10}\left(\frac{p_1}{p_2}\right)$$

where both powers p_1 and p_2 are measured in the same units, P is the logarithmic power ratio in decibels, and "$\log_{10}()$" represents the logarithm to the base 10 of the term in parentheses.[4]

---

[4] Throughout the text, I use lower-case letters for linear parameters and capital letters for logarithmic parameters for

Since antenna gain g is actually a power ratio, we can thus write it in logarithmic form as

$$G = 10Log_{10}\left(\frac{p_1}{p_2}\right) \qquad (2.2)$$

where gain G has units of decibels. Typical antenna gains for meteorological radars range from 20 to 45 dB. The gain of an isotropic radiator would be 0 dB (i.e., $p_1 = p_2$, so $p_1/p_2 = 1$).

Another important parameter of an antenna is its beamwidth. The beamwidth of an antenna is defined as the angular width of the antenna beam measured from the point where the power is exactly half what it is at the same range on the center of the beam axis. Figure 2.6 illustrates the beamwidth of an antenna.

Antenna gain and antenna beamwidth are related. One expression that can be used to calculate one from the other is (Battan, 1973)

$$g = \frac{\pi^2 k^2}{\theta \phi} \qquad (2.3)$$

where θ and ϕ are the horizontal and vertical beamwidths of the antenna, respectively, and both are measured in radians. k^2 depends upon the kind and shape of antenna. For circular reflectors, $k = 1$.

---

those parameters which are commonly expressed in both linear and logarithmic units of measure. See Appendix A for a more complete discussion of logarithmic measurements. See Appendix B for a discussion of error analysis using logarithmic parameters.

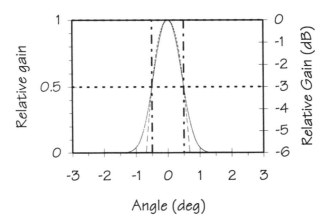

Figure 2.6 The relative gain of an ideal antenna as a function of angle. The antenna beamwidth of this antenna is 1°. The solid line is the linear antenna gain (left ordinate) while the dashed curve is the logarithmic gain (right ordinate). The beamwidth is the angular width where the power is exactly half of the maximum power. On the linear scale, this is at a relative power of 0.5 the maximum. On the logarithmic scale, the half-power point is 3 dB below the maximum. The dash-dot-dash curve represents a top-hat profile; it is physically unrealizable but what we would like a radar to have (left scale applies to the top-hat profile).

For circular reflectors, the horizontal and vertical beamwidths would be equal, giving

$$g = \frac{\pi^2}{\theta^2} \qquad (2.4)$$

For example, for an antenna with a 1° beamwidth, the gain would be

$$g = \frac{\pi^2}{\left(1° \frac{\pi}{180°}\right)^2}$$

$$= 32400$$

or, in logarithmic units,

$$G = 45.1 \text{ dB}$$

Notice that gain is independent of wavelength. *Any* circular parabolic radar antenna with a 1° beamwidth would have the same antenna gain at any frequency, according to Eq. 2.4.

The shape of the mainlobe is often approximated by a Gaussian shape (Probert-Jones, 1962). A Gaussian beam pattern can be written in an equation of the form

$$g = g_0 \, exp\left(-\frac{2\theta^2}{\theta_0^2}\right) \tag{2.5}$$

where g is the gain at any arbitrary angle θ from the center of the mainlobe axis, θ_0 is the beamwidth of the mainlobe, and g_0 is the maximum gain on the beam axis. If g_0 is set to 1, then g is the relative gain of the Gaussian beam pattern.

As mentioned, gain is frequently expressed logarithmically. One equation for doing this is (Doviak and Zrnic', 1993)

$$G = -16 \ln(2)\left(\frac{\theta}{\theta_o}\right) \tag{2.6}$$

where G is the relative logarithmic gain (in decibels) at any angular distance θ from the beam axis, and θ_o is the beamwidth of the pattern. By multiplying the right side of Eq. 2.6 by the maximum gain on the beam axis G_o, the gain G becomes the absolute gain in the direction of θ.

According to Eq. 2.6, a radar antenna's mainlobe beam pattern (measured in decibels) decrease in magnitude approximately, proportional to the square of the angular distance from the beam axis. If mainlobes really are Gaussian in shape, the decrease would be exactly as the square of the angular distance. Actual antennas differ slightly from this, however. For example, the National Center for Atmospheric Research CP2 S-band radar antenna had a mainlobe that decreased approximately to the 2.2 power of angular distance from the mainlobe.

Much as we would like to believe it, antennas are not perfect devices. The ideal antenna would direct all of the radar's energy into a single direction and none of it would go anywhere else. This is physically impossible. Even flashlights do not do this job perfectly. While most light from a flashlight does go in some preferred direction, some of the light can be seen well off to the sides of the brightest spot. Further, the illumination of the brightest spot is seldom uniform.

Real radar antennas are much like this. They will have a bright spot (called the mainlobe), but they will also transmit and receive energy off to the side of the mainlobe in what are called sidelobes. Further, the sidelobes exist in all directions away from the mainlobe and are different from one direction to another. One difference between radar antennas and flashlights is that some of the radar energy can actually go directly behind the antenna, forming a "backlobe".

The top-hat antenna gain pattern shown in Fig. 2.6 is for a perfect antenna while the Gaussian antenna beam pattern is a reasonable approximation to the mainlobe of real antennas. But in either case, the pattern on Fig. 2.6 has no sidelobes whatsoever. Real antennas have sidelobes, and sometimes very strong sidelobes. Let's examine some sidelobes from a couple of radar antennas. When we examine antenna beam patterns, we usually only do so in one direction at a time, either in azimuth or in elevation. Figure 2.7 shows the antenna beam pattern in the horizontal direction for the AN/CPS-9 X-band antenna used by the Air Force and others during the 1950's and '60's (Donaldson, 1964). This pattern is a smoothed, idealized fit to the real pattern. It shows that the simple, single lobe pattern of Fig. 2.6 is not a good approximation only near the center of the mainlobe. It does not represent the sidelobes at all!

Now let's look at the measured antenna beam pattern from a real antenna. Figure 2.8 shows part of the antenna beam pattern for the FL2 radar, an S-band (10-cm wavelength) radar operated by MIT Lincoln Laboratory (see Appendix D). This pattern was measured by placing a calibrated signal generator at some distance from the radar antenna and scanning the antenna slowly in azimuth through almost a full circle while receiving and recording the signal. The strong mainlobe is clearly evident when the antenna was aimed directly at the signal generator. Nearby sidelobes are also shown as moderately strong but narrow "spikes" on the pattern. Near 120° either side of the mainlobe are regions of stronger sidelobes. Notice that

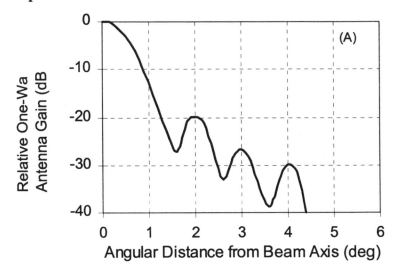

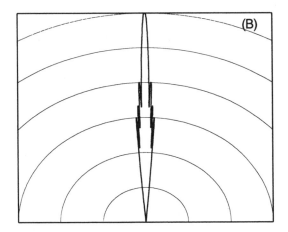

Figure 2.7 Modeled one-way antenna beam pattern for the CPS-9 X-band antenna, showing the mainlobe and the first three sidelobes. This antenna has a mainlobe with a 1° beamwidth. From Donaldson, 1964. (A) shows the beam patter as function of gain and angle. (B) shows the pattern in polar coordinates (10-dB contours). It shows that the sidelobes are really quite close to the mainlobe of the beam pattern

some power was even detected when the antenna was aimed in the opposite direction from the signal generator (i.e., at an azimuth of 180°).

As complex as the antenna beam pattern is that is shown in Fig. 2.8, it does not really portray the complexity of a complete antenna beam pattern. The pattern shown is, after all, a single slice through what is really a two-dimensional pattern. As an example of the complexity of a real pattern in *both* azimuth and elevation, Fig. 2.9 shows the antenna beam pattern for the CP2 X-band Cassegrain-feed antenna of the National Center for Atmospheric Research (Rinehart and Frush, 1983). This pattern was obtained by transporting the antenna to the antenna range of the National Bureau of Standards in Boulder, Colorado, an expensive and time-consuming activity. As can be seen in the figure, there are a number of sidelobes encircling the mainlobe, with each successive sidelobe ring generally being of weaker strength. Note that this pattern only covers 10° of azimuth and elevation, 5° either side of the center of the mainlobe. A complete antenna beam pattern ±180° in azimuth and elevation around the mainlobe would be even more complex.

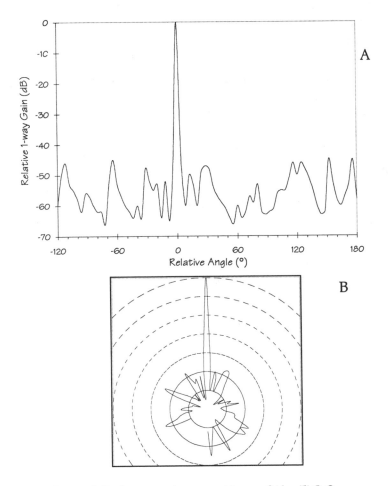

Figure 2.8 Antenna beam pattern of the FL2 S-band radar. This pattern was taken using vertical polarization and is through the center of the beam axis at 0° elevation angle. (A) shows the pattern in terms of gain and angle while (B) shows the pattern in polar coordinates (10-dB contours).

Transmit/Receive Switch

The transmit/receive (T/R) switch shown in Fig. 2.1 is a special switch added to the radar system to protect the receiver from the high power of the transmitter. Most radars transmit from a few thousand watts to more than 1 MW (1 MW = 10^6 watt) of power. Most radars are capable of receiving powers as small as 10^{-14} W or less. Because of this tremendous difference in power levels, if a transmitter sent much of its power into the receiver, it would very quickly burn up the very sensitive receiver.

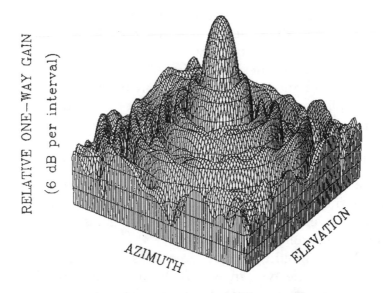

Figure 2.9 Antenna beam pattern of the NCAR CP2 X-band antenna. The elevation and azimuth angles extend about 5° either side of the mainlobe (0.1° per interval for both elevation and azimuth). The horizontal contours are at 6-dB intervals. From Rinehart and Frush, 1983.

In order to protect the receiver from this possibility, radar engineers have added the automatic switch (also called a duplexer or circulator) in the waveguide between the transmitter and the receiver. When the transmitter is turned on, the duplexer acts to direct the strong pulse of energy to the antenna and away from the receiver. As soon as the transmitter stops sending its signal, the duplexer switches so that the receiver is connected to the antenna while the transmitter is disconnected.

Transmit/receive switches do not respond instantaneously. As a result, there will be a short recovery time after the transmitter fires before the receiver is at full sensitivity. This recovery time depends upon the design of the T/R switch and the power of the transmitter, but is typically on the order of 10 μs; this corresponds to a distance of a couple of kilometers.

Receiver

The receiver is designed to detect and amplify the very weak signals received by the antenna. Radar receivers must be of very high quality because the signals that are detected are often very weak.

Most weather radar receivers are of the superheterodyne type in which the received signal is mixed with a reference signal at some frequency which is different from the transmitted frequency. Mixing the received signal with the reference frequency converts the signal to a much lower frequency (typically 30 to 60 MHz) which is more easily processed. This lower frequency (intermediate frequency or IF) is further amplified to the point that it can be used for displays and recording. The connection between the receiver and the indicator on Fig. 2.1 is usually coaxial cable;

since the frequency of this signal is much lower and the distances between devices at this point in the receiver are quite short.

Modern Doppler radars often contain two receivers. One receiver is designed to cover the wide range of reflectivities that are present in the real world, sometimes up to 8 or 10 orders of magnitude. Logarithmic receivers are used to handle this wide range of reflectivities. A logarithmic receiver produces an output which is proportional to the logarithm of the input power. Consequently, they can have a dynamic range (the difference from the weakest to the strongest power that it can handle) of 80 dB or more.

For Doppler velocity measurements, the receivers that are used are usually somewhat more sensitive to weak echoes but of much more limited dynamic range. Moderate and strong signals will overwhelm or saturate these receivers, but they still give correct velocity data. One consequence of using two receivers is that sometimes very weak echo will show up better on a display of velocity data than it does on the reflectivity display.

Displays

There are many ways to display radar data. The earliest and easiest display for radar data was to put it onto a simple oscilloscope where the horizontal axis was time and the vertical axis was signal strength or intensity; an oscilloscope used to display radar data is called an A-scope; it was the first display "named" when radar was first invented. Since electromagnetic radiation travels at the speed of light, the time base can also be a distance scale. The vertical scale can also be calibrated in power units. Figure 2.10 shows a schematic A-scope display.

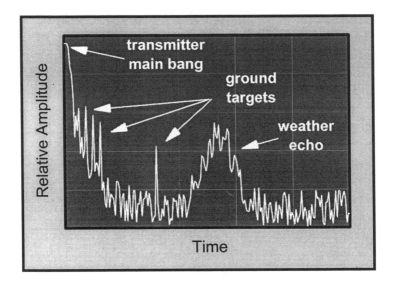

Figure 2.10 Schematic diagram of an A scope showing the transmitter pulse ("main-bang") leakage into the receiver, nearby ground targets, receiver noise, distant weather echo and an isolated point target.

One problem with the A-scope display is that it gives no direct information about where the radar antenna is pointing. Usually we want to know *where* the target is, not just its range and strength. In order to give position information, other displays had to be invented. Perhaps the most universal display for weather information is the plan position indicator (PPI). A PPI displays the radar data in a map-like format with the radar at the center. Distance is given by adding range marks (also called range rings) around the radar. The direction from the radar is shown by the position of the echo relative to the radar. Most radars put north at the top and have east to the right, south at the bottom and west to the left. Occasionally a radar will be aligned with magnetic north at the top. This alignment to magnetic north is convenient if the radar is used to direct the activities of aircraft (which fly using magnetic compasses).

Mobile radars used on ships typically have PPI displays that show a full 360° of coverage. They will usually put the heading of the ship at the top of the display so that what is ahead of the vessel will always be easily recognized.

Airborne radars are usually mounted in the front of the aircraft and have restricted antenna movement. That is, the antennas are not able to look around in all directions but can only scan from the left to the right of the aircraft's path by perhaps ±45° to ±60°. Consequently, aircraft radar displays do not show full-circle PPI's but only the directions scanned by the antennas. They still portray the information in range and azimuth, however. The radar display points in the direction the aircraft is flying.

Figure 2.11 shows a schematic example of a PPI display. The radar is located at the center of the display with range increasing outward to the maximum range of the radar (or display). This particular display shows radar targets as shaded. Near the radar is a large region of shading which could be either a storm echo over the radar or it could be nearby targets (often called ground clutter). Farther from the radar to the southwest are smaller targets. Again, these could be small weather echoes, but they are drawn to represent what aircraft echoes might look like (or ship echoes if the radar is located on a ship; in this case the center echo could be storm or sea clutter). To the north is a line of scattered storms; those with the black centers are the strongest.

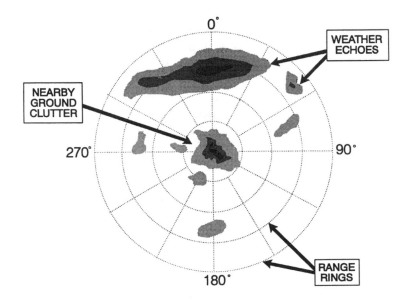

Figure 2.11 Schematic diagram of a PPI display showing nearby ground clutter and distant weather echoes. Echo intensity is illustrated by both contours and shading, where the darker the shading, the stronger the echo.

Modern technology has added a new dimension to radar displays -- color. Most modern radars have computer-generated displays which show not only the position of the radar echoes as on old-fashioned PPI's, they also show the intensity of the echo in (false) colors.[5]

[5] Some purists insist on specifying that color displays of radar data (or satellite data or other kinds of data) are "false" color. Obviously, storms are not green, yellow, or red. All the radar really detects is the power of the received signal or the velocity of the moving targets. It could display these in a single color as in the days of old, but by color coding intensities or velocities, it is much easier for a human observer to interpret storm intensity or motion fields. The colors

Older displays showed intensity by varying degrees of brightness. It was not possible to determine more than a few levels of intensity with these monochrome displays. Modern displays can easily show several levels of intensity quite clearly using color coding; some can show 15 or more different intensity levels. This is quite useful to meteorologists for quantitatively determining rainfall intensity levels, for example.

Another advantage of computer-generated displays is that they can put the radar anywhere, not just at the center of the display. Quite often the region of interest is off on one side of the radar only. By offsetting the display to be centered on this region, the same storms can be shown bigger than they could be on older displays. Another modern feature is the ability to magnify or zoom the image to fill the display area with a specific region of radar coverage. Again, this makes it possible to see the smaller-scale features of storms in much greater detail.

An example of a color display of real radar data in PPI format can be found in Color Fig. 1. It shows both the signal strength (radar reflectivity in units of dBZ; right side of the figure) and Doppler radial velocity data (left side) for the UND radar operating at South Roggen, Colorado. At the time this picture was taken, there were no weather echoes present. All of the echo shown is from ground targets, a few aircraft, and noise in the radar system. The strong echoes to the west (left) are

---

chosen are purely arbitrary and, in the one sense they are false, but it is not really necessary to carry this lexical usage too far. We will simply refer to them as colors, recognizing that they are man-made colors, not natural characteristics of real storms.

the Rocky Mountains west of Denver. Nearby ground clutter is also present, including a couple of small patches to the north at 30 to 45 km range. The velocity data show how fast the targets are moving. Except for a few aircraft in the area, nearly all of the echo shown has velocity near zero, evidence of the fact that it is from ground targets and are not moving. Other examples of PPI data are shown in the section of color figures.

Another kind of useful display for weather information is the range-height indicator (RHI). In this display the horizontal axis is again distance from the radar, but the vertical axis is height above the radar. Echoes are shown as bright or colored regions on the display. Color Fig. 4 (and others) shows an example of an RHI taken from the UND radar in Kansas City, Missouri, on 18 June 1989. The echo displayed is from a rapidly moving cold front approaching the radar from the northwest. The PPI in Color Fig. 3 shows the horizontal view of the same situation. It is often very helpful to have both the horizontal (PPI) and vertical (RHI) view of a storm to understand what is going on.

One problem with many RHI displays is that they exaggerate the vertical size of an echo. This is unfortunate because it gives a very distorted impression of what storms are really like. Convective storms are usually approximately spherical or even pancake shaped. Displaying them with vertical magnifications of 5 to 10 makes them look much taller than they are wide; such is not the case in nature. Color Figs. 4, 8, and 10 are not all exactly one-to-one in their vertical-to-horizontal scales. Be aware that many RHI's are far more distorted than those shown herein.

Signal Processor

The block diagram shown in Fig. 2.1 ends with the indicator. However, most modern radars have computer processors incorporated within the system which can do some very useful things. The color displays described above are examples of one use of these processors.

Research and some new operational radars (e.g., WSR-88D) are also equipped to process the radar data to detect various hazardous weather situations using what are called computer algorithms or simply algorithms. An algorithm is a specific set of instructions that the computer will execute to see if the storm contains specific features. If the characteristic the computer is looking for exists, it will indicate that the event is present. For example, it could sound an alarm if a tornado mesocyclone is detected.

The new radar now being used by the National Weather Service, the Federal Aviation Administration and the Department of Defense (NEXRAD, NEXt generation weather RADar, designated WSR-88D) is able to automatically detect such things as hail, tornadoes, and microbursts. Automatic warnings of these events are available for dissemination through appropriate channels to the public. See Chapter 11 for a more detailed discussion of the WSR-88D NEXRAD radars.

RADAR

SIMPLIFIED BLOCK
DIAGRAM OF A RADAR

REK/'91

Chapter 3

Electromagnetic Waves

Radio and radar both operate using electromagnetic radiation. Electromagnetic radiation, as its name suggests, has both electric and magnetic components, each component of which is like a magnetic wave and an electric wave vibrating at right angles to each other, and both are at right angles to the direction of propagation. Electromagnetic radiation always travels at the speed of light (where light, itself, is just a special form of electromagnetic radiation; it just happens to be at a frequency and wavelength which is detectable by our eyes).

One of the important characteristics of electromagnetic radiation is its frequency. Another is its wavelength. These are related through the equation

$$f = \frac{c}{\lambda} \tag{3.1}$$

where f is frequency in hertz (1 Hz = 1 cycle/second), c is the speed of light (often measured in m/s) and λ is wavelength (in meters when c and f are in the units specified).

Chapter 3

Electromagnetic spectrum

Electromagnetic radiation ranges from very low frequencies to very high frequencies. Figure 3.1 shows what is often called the electromagnetic spectrum and where various useful frequencies are located within it. Commercial electricity uses 50 (much of the world) and 60 Hz (North America) near the low end of the spectrum. Radio waves are at somewhat higher frequencies. Radar is located near the upper end of this figure. Infrared, visible, and ultraviolet light are to the right of the figure. X-rays and gamma waves are even higher frequency forms of electromagnetic radiation.

The frequencies used by radars range from perhaps 100 MHz through 100 GHz. Certain frequencies have been so frequently used for radar that it has been found convenient to designate them by letters. Table 3.1 lists the bands commonly used with radar along with their frequencies and wavelengths. With one exception, the designations given are those that have been in use for the past 50 years or so. The exception is the last entry in the table. Since the highest frequencies were the last to be utilized, they were the

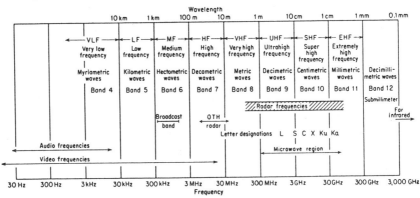

Figure 3.1 Electromagnetic spectrum. From Skolnik, 1980, Introduction to Radar Systems, with permission from McGraw-Hill, Inc.

last to be given band designations. Until recently, the "millimeter" band was simply called the "mm" band. Recently I have seen it designated as the W-band (Sassen and Liao, 1996). A new set of letter designations was proposed some time ago in an attempt to get a logical system into this categorization, but most radar meteorologists and other radar people, however, continue to use the old familiar band letters.

Table 3.1   Radar bands and the corresponding frequency bands and approximate wavelengths.

Band Designation	Nominal Frequency	Nominal Wavelength
HF	3-30 MHz	100-10 m
VHF	30-300 MHz	10-1 m
UHF	300-1000 MHz	1-0.3 m
L	1-2 GHz	30-15 cm
S	2-4 GHz	15-8 cm
C	4-8 GHz	8-4 cm
X	8-12 GHz	4-2.5 cm
K_u	12-18 GHz	2.5-1.7 cm
K	18-27 GHz	1.7-1.2 cm
K_a	27-40 GHz	1.2-0.75 cm
mm *or* W	40-300 GHz	7.5-1 mm

Refractive Index

The speed of electromagnetic radiation depends upon the material through which it is traveling. In a vacuum such as the nearly empty space between the sun and earth, for example, light travels at a speed of 299 792 458 ±6 m/s, according to the National Bureau of Standards (Cohen and Taylor, 1987).

When electromagnetic radiation travels through air or other materials, it travels slightly slower than in a vacuum. The ratio of the speed of light in a vacuum to the speed of light in a medium is called the refractive index of the medium and is defined mathematically as

$$n = \frac{c}{u} \tag{3.2}$$

where c is the speed of light in a vacuum, u is the speed of light in the medium and n is the refractive index. Since c is always greater than or equal to u, n is always greater than or equal to 1. Note that n is a unitless parameter.

Actually, the refractive index of electromagnetic radiation has two components. The one described above is the simple, real component of the complex refractive index m which is given by

$$m = n - ik \tag{3.3}$$

where $i = \sqrt{-1}$ and k is related to the absorption coefficient of the medium. For a perfect dielectric (nonconductor), k is equal to zero. For many purposes, all we need be concerned about is the real component n.

In the atmosphere near sea level, the refractive index of air is approximately 1.0003. This means that electromagnetic radiation travels approximately 0.03% slower there than in a vacuum. Obviously, the refractive index must decrease from 1.0003 near the surface of the earth to 1.0000 at the top of the earth's atmosphere. Usually there is a gradual decrease in this parameter with increasing height, but there can be more abrupt changes in some layers in the atmosphere occasionally.

Refractivity

Since the important part of the refractive index is in the fourth, fifth, and sixth decimal places (i.e., 321 of $n =$ 1.000321, for example), scientists have found it convenient to define another parameter which is easier to work with. This kind of modification is often done in science. We almost always find it easier to deal with numbers between 0 and 1000 or so than to deal with very small or very large numbers or numbers that differ only slightly from some constant value. In the case of the index of refraction, by subtracting 1 and multiplying the results by 1 000 000 (i.e., 10^6), we get a very convenient number; this new parameter is called refractivity N and is sometimes said to be measured in N-units. For the example given above where $n = 1.0003$, $N = 300$. In equation form, N is defined as follows:

$$N = (n-1)\,10^6 \qquad\qquad (3.4)$$

The refractive index of the atmosphere has been found to depend upon atmospheric pressure, temperature and vapor pressure. It also depends upon the number of free electrons present. However, in the troposphere there are insufficient free electrons to be important. The effect of free electrons is important high in the atmosphere. In fact, it is the detection of variations in free electron concentrations that allows wind profiling radars to detect winds in the upper stratosphere and mesosphere (i.e., the ionosphere). We will ignore this effect within the troposphere

The equation relating refractivity to atmospheric variables is

$$N = \frac{77.6}{T}\left(P + 4810\frac{e}{T}\right) - 4.03 \cdot 10^7 \frac{N_e}{f^2} \qquad (3.5)$$

where T is atmospheric temperature (in kelvins, i.e., degrees above absolute zero on the Kelvin scale); P is atmospheric pressure (in millibars [mb] or hectopascals [hPa]); e is the vapor pressure of the moist air (in mb or hPa); N_e is the number density of free electrons per m^3; and f is the operating frequency of the radar in Hz. The numerical constants were determined empirically. The right-most term is important only in the ionosphere; we will ignore this term henceforth.

In the troposphere, refractivity is determined from temperature, pressure and vapor pressure. These are available from soundings of the atmosphere which are made twice a day at many radiosonde stations throughout the world. From sounding data we can calculate N for each level in the atmosphere; from N and height measurements, we can get the gradient of N with height $\delta N/\delta H$ where δN is the change in N over a given change in height δH. As mentioned, N normally decreases with increasing height, so $\delta N/\delta H$ is normally negative.

As an example of the effects of atmospheric conditions on radar propagation, consider the sounding shown in Fig. 3.2. This sounding (Bangkok, Thailand) has a warm moist lower layer with a slight inversion near the surface. From the temperature, dew-point temperature and pressure on this sounding, we can calculate the refractive index n and refractivity N for the sounding.

Figure 3.3 shows the refractivity as a function of height for the same sounding as shown in Fig. 3.2. The overall trend is that refractivity decreases with altitude, but at a gradually decreasing rate. There are a couple of models that are applied to atmospheric refractivity soundings. We will see shortly that one model suggests that refractivity decreases at a constant rate, the accepted "standard" value being at a rate of -39 N-units/km. Another model suggests that the refractivity decreases at a logarithmic rate. The thin smooth line on Fig. 3.3 is a logarithmic curve fit to the data (using least-squares statistical techniques). It fits the general features of the actual sounding quite well.

Radar propagation is more dependent upon the *gradient* of refractivity rather than the absolute value of refractivity at any point. Figure 3.4 shows the gradient of refractivity as a function of altitude for the data shown in Figs. 3.2 and 3.3. Clearly, the sounding is *not* standard! There are a couple of shallow layers where the gradient is more negative than -39 N-units/km, but much of the sounding has a gradient greater than this. Near the surface, the inversion produces a strong region of gradient that is actually positive (+81 N-units/km). This would be a layer of subrefraction.

Based on a sample of one (a dangerously small sample size!) it appears that the logarithmic change in refractivity with altitude is a much better fit to the real atmosphere than the linear model. Nevertheless, the linear model is still quite useful and will be used later in this chapter as the easiest way to handle radar propagation in the troposphere.

Another way to determine the refractive index of the atmosphere is to measure it with an instrument called a refractometer. Microwave refractometers, for example, are devices which have a chamber through which environmental air is pumped. The resonant frequency of this chamber depends upon the size of the chamber and the refractive index of the air in it.

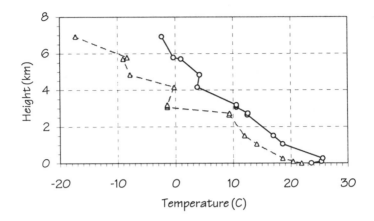

Figure 3.2 Sounding of temperature (right curve) and dew-point temperature (left curve) for Bangkok, Thailand, November 1996.

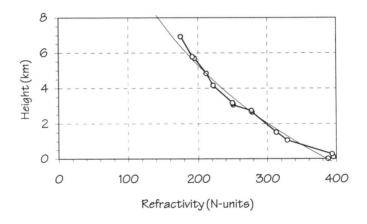

Figure 3.3 Refractivity N as a function of altitude for the sounding shown in Fig. 3.2. The thin curve is an exponential fit to the actual profile.

One of the consequences of having a different refractive index at different places in the atmosphere is that the electromagnetic radiation will travel at different speeds. In the lower portions of the troposphere, the atmosphere tends to be stratified into horizontal layers much of the time. Changes in refraction are usually much stronger in vertical directions than in horizontal directions. Thus, a wave of electromagnetic radiation which is traveling horizontally at one point will travel faster at one level and slower at another. The wave front which was originally perfectly straight up and down will gradually bend one way or the other. The direction the wave front bends depends upon whether the lowest refractivity (hence the fastest speed) is on the top or the bottom.

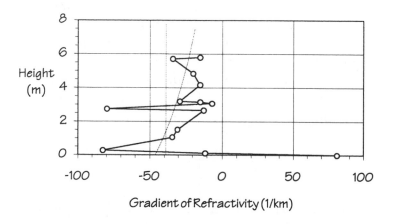

Figure 3.4 Gradient of refractivity as a function of altitude. "Standard" refraction corresponds to the vertical dashed line at -39 N-units/km. The curved smooth line is based on the exponential fit to the refractivity on Fig. 3.3.

Chapter 3

Under normal atmospheric conditions, N is largest near the ground and decreases with height. This means that radar waves will travel faster aloft than near the surface. This bends the waves in a downward direction relative to the earth's surface (i.e., relative to the horizontal).

Snell's Law

It is often more convenient to talk about radar rays rather than radar waves. Rays are lines along which waves travel and are drawn perpendicular to the wave fronts. If the wave bends, the rays bend correspondingly. Rays are especially convenient in optics to show how light travels through lenses. They are also convenient in determining the paths radar waves will follow in the atmosphere.

By knowing the refractive index in the atmosphere at each level, we can predict the path radar waves will follow. One way to predict this is through the use of Snell's law. Snell's law predicts the bending that light or electromagnetic radiation will undergo when it travels from one medium to another, each having its own refractive index. Snell's law can be written a number of ways. One way is as follows:

$$\frac{\sin i}{\sin r} = \frac{u_i}{u_r} = \frac{n_r}{n_i} \tag{3.6}$$

where i is the angle of incidence and r is the angle of refraction. u_i and u_r are the speeds of electromagnetic radiation and n_i and n_r are the refractive indices in the first and second layers, respectively.

Using Snell's law, if we know the starting angle of a radar wave at one place, the refractive index at that point, and the refractive index at the next higher level in the atmosphere, we can calculate the angle the air will have in the second layer. By doing this for each layer in the atmosphere, it is possible to predict the path that rays will follow anywhere in the atmosphere. Calculating ray paths in the atmosphere is useful for determining how the rays will behave in the real world.

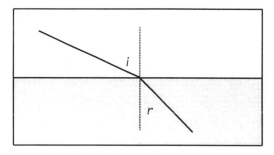

Figure 3.5 Illustration of a ray of electromagnetic radiation passing from one medium to another. The more dense medium is on the bottom. The angles of incidence and refraction are shown.

If there were no atmosphere on earth or if the atmosphere were perfectly uniform everywhere, radar rays would travel in straight lines. If a ray were to leave a radar traveling perfectly horizontally and continued in a straight path, the curvature of the earth beneath the ray would gradually cause the ray to be higher and higher above the earth farther and farther from the radar. This effect is sometimes called the earth's curvature effect. In this example the ray has gone perfectly straight; the earth has curved out from underneath it. If, on the other hand, an observer had left the radar and traveled underneath the radar's beam out some distance, they would have found that the beam was still getting higher above the ground at increasing distances. Relative to the earth, the radar wave would appear to bend; that is, the radar ray would have curvature relative to the earth.

Curvature

Mathematically, what is curvature? Curvature is defined as "the rate of change in the deviation of a given arc from any tangent to it." Stated another way, it is the angular rate of change necessary to follow a curved path. Mathematically, we can write curvature as $\delta\theta/\delta S$ where $\delta\theta$ is the change in angle experienced over a distance δS.

Another definition of curvature is that it is the reciprocal of the radius. Think about a circle. The distance around a circle of radius R is the circumference given by $\delta S = 2\pi R$. In traveling around a circle, the angular distance $\delta\theta$ traversed is 2π radians. The curvature C of a circle is thus the angular distance traversed divided by the linear distance traversed, or

$$C = \frac{\delta\theta}{\delta S} = \frac{2\pi}{2\pi R} = \frac{1}{R} \qquad (3.7)$$

Curvature has units of reciprocal length, e.g., 1/km or km^{-1}.

For a radar ray traveling relative to the earth when there is a non-uniform atmosphere present, the ray will bend more or less relative to the earth, depending upon how much the refractive index changes with height. Consequently, the curvature of a radar ray relative to the earth's surface is given by (Battan, 1973)

$$\frac{\delta\theta}{\delta S} = \frac{1}{R} + \frac{\delta n}{\delta H} \qquad (3.8)$$

It is sometimes convenient to think of the radar rays traveling in straight lines instead of the actual curved paths they do follow. We can accomplish this by creating a fictitious earth whose radius is different from the true earth's radius. This effective earth's radius R' is given by

$$\frac{1}{R'} = \frac{1}{R} + \frac{\delta n}{\delta H} \qquad (3.9)$$

Consider the case of a radar ray bending exactly the same as the earth. In this case the curvature *relative to the earth's surface* is zero. From Eq. 3.8 we can write

$$\frac{\delta\theta}{\delta S} = \frac{1}{R} + \frac{\delta n}{\delta H} = 0 \qquad (3.10)$$

so

$$\frac{1}{R} = -\frac{\delta n}{\delta H} \qquad (3.11)$$

Since the radius of the earth $R = 6374$ km, then the refractive index gradient $\delta n/\delta H$ needed for a ray to follow the earth's surface is $-1.57 \cdot 10^{-4}$ km^{-1} or, in N-units, this is -157 N-unit/km.

Using these various relationships between curvature, earth's radius, effective earth's radius, and refractive index gradient, it is possible to calculate the actual path a radar ray will follow in real atmospheric conditions. Most such calculations assume that "standard refraction" conditions apply. Standard refraction is when $\delta N/\delta H$ = -39 N-units/km. This is the value needed to have straight radar rays in the standard atmosphere when using the normally-accepted value for the effective earth radius R'. With $\delta N/\delta H$ having the value given above, we can calculate R' from the equation defining R'. To do that, however, we have to convert $\delta N/\delta H$ into $\delta n/\delta H$. Doing this gives

$$\frac{1}{R'} = \frac{1}{6374 km} - \frac{39 \bullet 10^{-6}}{km} = \frac{1.179 \bullet 10^{-4}}{km}$$

or R' = 8483 km $\cong$ 1.3 $R \cong$ 4/3 R, the approximate value for standard conditions.

The actual value of the effective earth radius R' under any given set of temperature, pressure, and humidity conditions can be calculated from sounding information. R' varies from perhaps 1.1R to 1.6R, depending on conditions. The value usually used (4/3R) applies to most normal conditions fairly well. Be aware, however, that it is only an approximation and errors will be made if it is applied blindly to all conditions.

Super Refraction

When the downward bending of radar waves is stronger than normal, we call this superrefraction. It occurs, for example, when the temperature increases with altitude (i.e., when an inversion is present). Superrefraction can allow a radar to detect ground targets to much longer distance than under "normal" conditions. Since nocturnal inversions occur frequently in many parts of the country, extended range detection of ground targets is also quite common, especially at night and in the early morning hours. It also occurs under other conditions such as when the radar is looking under a thunderstorm. The condition of extended range of detection of ground targets is called anomalous propagation or anoprop (AP). AP is detected by most ground-based radar sites, at least occasionally.

If the refraction of the radiation is strong enough, the radar waves can be trapped in a layer of the atmosphere. When this happens, we call it ducting. Ducting occurs when $\delta N / \delta H <= -157$ N-units/km. Most radar ducts are close to the earth's surface, but under some conditions radar ducts can exist above the surface of the earth. For an airborne radar flying within a radar duct, it would be possible for it's signal to be trapped within the duct so that targets above or below the duct would not be detectable. Fortunately, ducting is most prevalent under fair weather conditions, so it should not interfere with the detection of most meteorological echoes.

Finally, ducting also depends upon the wavelength of the radar. The longer the wavelength, the deeper the layer has to be before ducting can take place. Thus, shorter wavelength radars suffer from ducting more than longer wavelength radars do.

Subrefraction

Sometimes radar waves are not bent downward as much as usual or, under more extreme conditions, may even be bent upward. This condition is called subrefraction. While less common, it can also cause problems with radar detection of targets.

The military is particularly concerned about knowing how radar waves will propagate in the atmosphere. They want to know when and where sub- and superrefraction will occur. That way they can be alert to the possibility of enemy aircraft or missiles coming into a region and hiding in layers where they might be undetectable.

Standard Refraction

When standard refraction applies, the height of the radar beam can be given by the following equation:

$$H = \sqrt{r^2 + R'^2 + 2rR' \sin\phi} - R' + H_O \qquad (3.12)$$

where r is the range from the radar to the point of interest, ϕ is the elevation angle of the radar beam, H_O is the height of the radar antenna, $R' = 4/3\ R$, and R is the earth's radius. Any consistent set of units can be used with this equation. For metric measurements, $R = 6374$ km.

Actually, since the earth is not a true sphere but more of an oblate spheroid, the actual radius that should be used depends upon the latitude where the radar is located. For most practical purposes, however, the value given here is accurate enough.

Figure 3.6 gives height as a function of range and elevation angle for standard refraction conditions and is a useful way to determine the height of a radar beam. If the radar is located above sea level, its height must be added to the height determined from the graph to give heights above mean sea level.

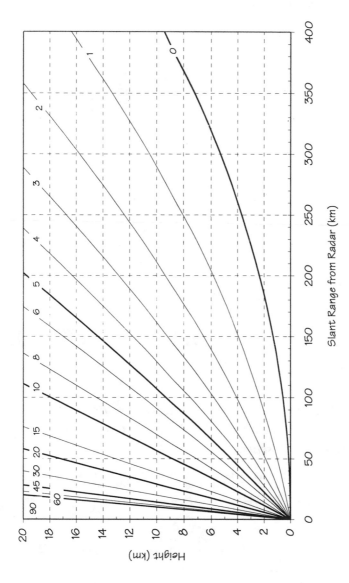

Figure 3.6   Range-height diagram. Numbers on each curve are elevation angles in degrees.

Forecast made in 1974:

Real-life observation and verification made in 1987, courtesy of Jim Fankhauser, National Center for Atmospheric Research:

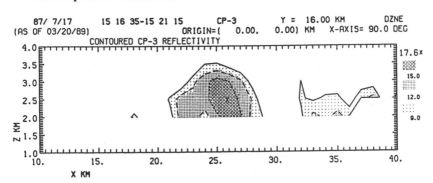

Chapter 4

Radar Equation for Point Targets

Radar is often used simply to show the locations of storms near a radar. But most radars are capable of not only detecting storms, they are also capable of measuring the strength of the returned power which in turn can be used to estimate rain rate and other parameters of the storms.

In order to use radar quantitatively, we must know the values of certain radar parameters. In this chapter we will discuss some of the theory behind the quantitative use of radar for the detection of isolated, point targets. In the next chapter we will build upon this information and cover the detection of beam-filling meteorological targets.

Point target radar equation

When a radar transmits a pulse of energy, the energy is directed into space by the antenna. Let's consider first an isotropic antenna. The power radiated moves away from the antenna at the speed of light, forming a spherically expanding shell of energy. The area covered by a single, expanding pulse of energy is equal to the area on the surface of a sphere at the corresponding distance, i.e.,

$$\text{area} = 4 \pi r^2 \qquad (4.1)$$

where r is the range from the radar (the radius of the sphere). The power density S, i.e., power per unit area, is simply the transmitted power divided by this area. Thus,

$$S = \frac{P_t}{4 \pi r^2} \qquad (4.2)$$

Figure 4.1a illustrates the power radiated from a point covering a sphere of radius r.

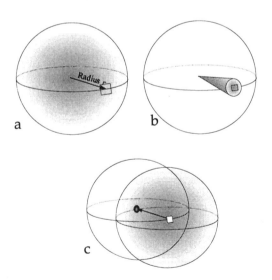

Figure 4.1 a) Power transmitted by an isotropic antenna expands to cover a sphere of radius r. b) Using an antenna, the power at a point on the beam axis is increased. c) The power

intercepted by an area A_σ is reradiated isotropically in all directions.

When a real antenna is used, the amount of power along the center of the beam axis at some distance is greater than it would be if an isotropic radiator is used. This increase in power is simply the gain times the power that would have been there if an isotropic antenna had been used. But now *more* power will be on the center of the beam axis while *less* power will occur in other directions. The radar still transmits the same amount of power, however, so the average power density would remain constant over the entire sphere.

Since we are now interested in what happens along the beam axis, we can introduce a target there that has an area A_σ. Thus, the power intercepted by the target will be given by

$$p_\sigma = \frac{p_t g A_\sigma}{4\pi r^2} \qquad (4.3)$$

where subscript σ represents the target. In this equation, the linear value of antenna gain is used, not the logarithmic value.

For most targets detected by a radar, the power intercepted is reradiated isotropically back into space (Fig. 4.1c). While some targets may radiate stronger in some direction than another, we can ignore this for the time being. Some targets may also absorb some of the incident energy, converting it into internal heat. This, too, will be ignored.

Chapter 4

When the target reradiates its energy, some of the energy will be received back at the radar. The amount of energy detected by the radar will be

$$p_r = \frac{p_\sigma A_e}{4\pi r^2}$$

$$= \frac{p_t g A_\sigma A_e}{(4\pi)^2 r^4} \qquad (4.4)$$

where A_e is the effective area of the receiving antenna.

Now, the effective area of an antenna A_e can be expressed in terms of the gain of the antenna and the wavelength λ of the radar. This is given by

$$A_e = \frac{g \lambda^2}{4\pi} \qquad (4.5)$$

We can substitute this expression into our radar equation, giving

$$p_r = \frac{p_t g^2 \lambda^2 A_\sigma}{64\pi^3 r^4} \qquad (4.6)$$

There is only one more refinement that needs to be made to this equation. This relates to the area of the target A_σ. The physical size of the target (which is also the size the target appears to the human eye) is not necessarily the size the target appears to the radar. To overcome this problem, we define a new parameter called the backscattering cross-sectional area of the target and give it the symbol σ; we can substitute this parameter for the area A_σ. Thus, the final form of the radar equation for a point target located on the center of the antenna beam pattern is

$$p_r = \frac{p_t \, g^2 \, \lambda^2 \, \sigma}{64 \, \pi^3 \, r^4} \qquad (4.7)$$

The backscattering cross-sectional area σ of a target is a function not only of the size, shape and kind of matter making up the target but also of the wavelength of the radar viewing it. Unfortunately, the backscattering cross-sectional area cannot always be calculated analytically, especially for complex targets. Fortunately for radar meteorologists, the shapes of many important targets are relatively simple. Most hydrometeors are approximately spheres, so let us consider spheres for a while.

Spherical Targets

When a sphere is large compared to the wavelength of the radar, the backscattering cross-sectional area of the target is equal to the geometric area. That is,

$$\sigma = \pi r^2 \qquad (4.8)$$

"Large" is usually interpreted to mean $D/\lambda > 10$ (although some specify $D/\lambda > 16$ for this condition to apply), where D is the diameter of the sphere and λ is wavelength.

When the size of a sphere is small compared to the wavelength of the radar, the sphere is in what is called the Rayleigh region. "Small" is usually interpreted to mean $D/\lambda < 0.1$ (although some specify $D/\lambda < 1/16$). In the Rayleigh region the backscattering cross-sectional area of a sphere is proportional to the sixth power of the diameter. Thus it is quite simple to calculate the return that can be expected from spheres in the Rayleigh region. The expression to calculate σ for a sphere is given by (Battan, 1973)

$$\sigma = \frac{\pi^5 |K|^2 D^6}{\lambda^4} \tag{4.9}$$

where $|K|^2$ is a parameter related to the complex index of refraction of the material. This will be covered in more detail in the next chapter.

Many meteorological targets really are small compared to the wavelength of a radar, so the Rayleigh region is an important part of meteorological radar use. And certainly there are a lot of targets which are large compared to the wavelength. But there is still the important intermediate region. In this region it is much more difficult to calculate the return that can be expected from a spherical target.

Mie, in 1908, determined the analytical expressions needed to calculate the backscattering cross-sectional area of spheres of all diameters. Figure 4.2 shows the normalized backscattering cross-sectional area of spherical targets as a function of the relative size of the target expressed as circumference/wavelength. By normalizing both axes this way, the graph becomes universally useful for all wavelengths and all diameter spheres. In fact, it applies just as well to optical wavelengths as it does to radar wavelengths.

The region to the left of the figure is the Rayleigh region. A careful check of the slope in this region will show that it rises only four orders of magnitude on the ordinate (vertical direction) for every one order of magnitude on the abscissa. The Rayleigh relationship suggests an order of magnitude change in diameter should result in six orders of magnitude change in σ, this is the diameter-to-the-sixth relationship mentioned above. The apparent "discrepancy" is caused by the fact that the ordinate is the backscattering cross-sectional area divided by the geometric area.

The optical region is on the right side of the diagram. As the size of the target gets larger and larger, the backscattering cross-sectional area of a target approaches the geometric area of the target. Between these two regions is the Mie or resonant region. Here σ can actually decrease as the size increases for certain sized particles. As will be seen later, this characteristic can be used to detect the presence of hail in some storms by using *two* radars which operate at different wavelengths to look at the same region in space simultaneously.

There are a number of targets which can be considered point targets for radars. We have already discussed spheres. However, to be a point target in the sense we have been discussing, we must have only a single target in the radar's sample volume. This is certainly not the case for many weather echoes. As the old saying goes, "when it rains, it pours!" A single radar sample volume of a thunderstorm might contain billions of raindrops (and even more cloud droplets). So rain and clouds are not considered point targets.

Not all point targets detectable by radar are spheres. Many, in fact, are so complex that they defy analytical solution. But these additional point targets are worth considering in some detail. So, in the following sections we will consider some other types of targets that can be detected by meteorological (and other) radars, starting, perhaps ironically, with spheres again

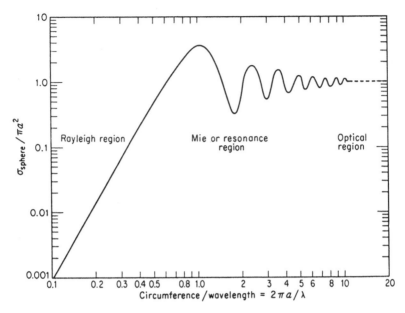

Figure 4.2 Normalized backscattering cross-sectional area of a sphere as a function of circumference normalized by radar wavelength λ. a = radius. From Skolnik, 1980, Introduction to Radar Systems, with permission of McGraw-Hill, Inc.

Standard targets

It is occasionally useful to aim the radar at a target with precisely known characteristics. Such targets are sometimes called *standard targets*.

Spheres

Spheres are useful as standard targets because they have the same backscattering cross-sectional area no matter what direction they are from the radar. Spheres can be tied to balloons and released and tracked by a radar or they can be tied beneath a balloon which is tethered to a fixed point on the ground. In either case, by measuring the power received from the sphere by the radar and knowing the size of the sphere and its distance from the radar, we can get a measure of the antenna gain of the radar system. This is done by solving the radar equation for a point target for antenna gain; all other parameters in the equation should be known or measurable at the radar site.

The magnitude of the backscattering cross-sectional area of a sphere can be obtained from Fig. 4.1 (Skolnik, 1980). If the sphere is large compared to the wavelength of the radar, then its backscattering area is the same as its geometric area, i.e.,

$$\sigma = \pi r^2 \qquad (4.10)$$

where r is the radius of the sphere.

Flat-plate reflectors, dihedrals, and trihedrals

Another class of standard targets is related to a flat-plate reflector. Flat-plate reflectors only work as intended when they are oriented such that they are perpendicular to the radar beam. If a flat-plate reflector is folded such that one side makes a 90° angle with the other side, it is called a dihedral reflector; to work correctly, a dihedral reflector must be oriented so that the folded axis is perpendicular to the radar beam. By putting three mutually perpendicular surfaces together, a trihedral or corner reflector is formed. Corner reflectors have the advantage that they do not need to be aimed toward the radar with great accuracy. Because of the three reflections that can take place, the reflected radar signal will always return directly along the path of the incident signal. When properly oriented, all three kinds of flat-plate type reflectors give very strong returns and can be used to measure the antenna gain of a radar.

The backscattering cross-sectional area of a flat plate reflector is given by (Levanon, 1988)

$$\sigma = \frac{4\pi A^2}{\lambda^2} \qquad (4.11)$$

where A is the geometric area of the target from the perspective of the radar, and λ is wavelength. If a square flat (unfolded) plate is normal to the antenna signal, $area = l^2$ where l is the length of one side of the square. If a dihedral or trihedral is used, then the geometric area as seen from the radar must be used. For example, the return from a trihedral corner reflector is given by

$$\sigma = \frac{4\pi\left(0.289 l^2\right)^2}{\lambda^2} \qquad (4.12)$$

where l is the length of one of the three sides of the reflector.

In all three cases with flat plate reflectors, notice that the size the target appears to the radar is usually much larger than its true geometric size. For example, a square target 1 m on a side would have a geometric area of exactly 1 m². A 10-cm wavelength radar, however, would receive a signal that would give a backscattering cross-sectional area of 1257 m²; that is 31 dB stronger than its geometric area. This enhanced return can be important for certain buildings as will be seen shortly.

Birds

Birds can be detected by many radars. There have been numerous studies into the detectability of birds by radar. Radar, in fact, is used by ornithologists to monitor the migratory habits of birds in various parts of the world.

As a first approximation, a bird can be regarded as a sphere of water whose mass is the same as that of the bird. This approach ignores the wings, tail, neck and head, and legs which might extend far from the body of the bird. Nevertheless, it is sufficiently useful that it can be used to predict how an individual bird might appear on a radar to within several decibels.

One of the problems related to detecting individual birds with radar is that they are point targets, and the power received from a point target is inversely proportional to the range to the fourth power. Thus, power decreases very quickly with increasing range. Consequently, the maximum range of detection of a bird is frequently only a few miles with some radars. Another problem is that they are relatively small targets. Birds such as starlings and pigeons have radar cross sections on the order of 10 to 20 cm^2 while herring gulls and mallard ducks have cross sections on the order of 80 to 90 cm^2 at S band.

Aircraft

The detection of aircraft was one of the primary motivations driving the development of radar during the 1930's and '40's and continues today as one of the most important uses of radar.

For most practical radars and aircraft, a single aircraft in a radar sample volume will be a point target. The return from an individual aircraft, however, is not simple but depends critically upon the relative orientation of the aircraft and the radar. Figure 4.2 (Skolnik, 1980) shows the measured cross section of one kind of aircraft; the return from other aircraft would be similar, but no two types of aircraft would have exactly the same "signature."

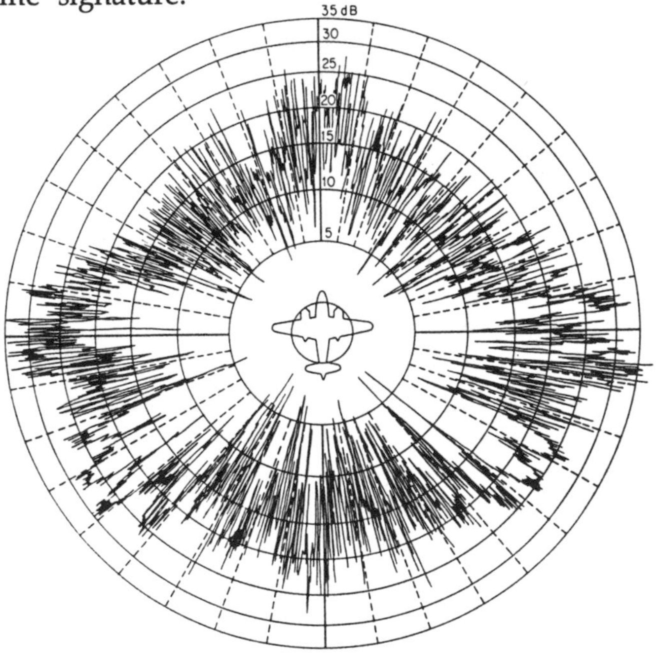

Figure 4.3 Experimentally measured cross-section of a B-26 two-engine bomber at 10-cm wavelength as a function of azimuth angle. From Skolnik, 1980, Introduction to Radar Systems, with permission of McGraw-Hill, Inc.

The distance to which a specific aircraft can be detected by a radar depends upon both the radar and the aircraft. The point target radar equation applies and states that the received power is inversely proportional to the distance to the target range to the fourth power.

So, signals from aircraft also drop off rapidly with range. However, since they are so much larger than birds, aircraft can usually be detected to moderately long ranges.

As an example, let us consider the detection of an aircraft with the ASR-9 air-traffic control radar used by the FAA (see Appendix D). Its transmitter power is 1.1 MW, its minimum detectable power is on the order of -114 dBm, it has an antenna gain of 34 dB. For the aircraft shown in Fig. 4.2, let us calculate the maximum range of detection. Assume that the radar cross section of the B-26 averages about 20 dBsm, where the units are decibels relative to a target of 1 m^2 backscattering cross sectional area. By rearranging the point target radar equation (Eq. 4.7) solving for range, we get:

$$r^4 = \frac{p_t\, g^2\, \lambda^2\, \sigma}{64\, \pi^3\, p_r}$$

Substituting in values and converting units gives

$$r^4 = \frac{1.1\,\text{MW}\; \left[10^{(34/10)}\right]^2 (0.107\,\text{m})^2\; 10^{(20/10)}\,\text{m}^2}{64\,\pi^3\; 10^{(-114/10)}\,\text{mW}\; 10^{-9}\,\text{MW/mW}}$$

so

$$r = 1001\ \text{km}$$

$$= 541\ \text{n mi}$$

This is a sufficiently long range that this aircraft should easily be detectable with the ASR-8 radar to very long ranges -- provided the target is on the center of the axis of the antenna beam pattern of the radar. This is not always the case, especially at long range where the earth's curvature effect puts the radar beam far above the surface (see Fig. 3.6).

Buildings

Many radars operate near or within radar line-of-sight of buildings. Individual buildings can also act as point targets to a radar. As a first approximation, the geometric area of a building should be about the same as the radar backscattering cross section; this assumes that the buildings will be in the "optical" region and somewhat irregularly shaped.

Many buildings, however, are built with lots of exterior right angles that can reflect a radar signal exactly the same way a dihedral corner reflector does. Consequently, just as with corner reflectors, the return from such buildings can be much stronger than the same sized building which has no such reflecting surfaces. In a similar way, rooms within buildings can act as trihedral corner reflectors, especially when they have lots of glass area and when metal construction is used. Thus, individual buildings can return large amounts of power to a radar.

Water towers and radio towers

One of the favorite point targets of many radar meteorologists is a well-chosen radio tower or water tower. It is useful to have such a target for each radar because they can be used as secondary standard targets. Towers have several useful properties: 1) They stick up high enough that they can be detected without contamination from other ground targets nearby. Sometimes the top of the tower can be detected while the base is invisible. 2) They are always at the same range and azimuth, so they can be used to verify the radar is displaying things in the correct location. 3) They should have a constant backscattering cross-sectional area, so they can be used to verify overall system sensitivity. This combination of attributes makes it nice to have at least one target you can detect to check the overall health and alignment of the radar being used. I would strongly recommend that you adopt a tower or two at each radar site.

Water towers sometimes have one additional attractive feature. Some water towers have relatively simple geometric shapes. An example is the kind of water tower that has almost a spherical tank atop a slender support tube. If such a tower is near a radar, it may be possible to get the dimensions of the tank from the city engineer, for example, and use it to calculate the backscattering cross-sectional are (which, for a large object like a water tank) should be the same as the geometric cross-sectional area. As such, the tower could conceptually be used as a standard target to measure antenna gain from Eq. 4.7.

Distributed point targets

This chapter has examined individual point targets, either aloft or at the ground. In Chapter 5 we will examine beam-filling meteorological targets. But there is at least one kind of target that is neither. This is a collection of point targets distributed over an area at the ground.

One of the significant differences between buildings (and other ground-based targets) and true point targets (e.g., birds and aircraft) is that buildings and trees often occur in clusters so that more than one target will be present within the same pulse volume.

The second significant difference is that buildings and trees are confined to the earth's surface (i.e., on a two-dimensional plain) rather than being distributed in three dimensions. As such, it is really more appropriate to consider the combined effects of "distributed point targets" rather than individual point targets.

One way to quantify the effects of several targets within the same pulse volume is to add up all the individual returns from each target and normalize (i.e., divide) by the area over which the targets are distributed;

$$\sigma_O = \frac{\sum \sigma_i}{area} \qquad (4.13)$$

where the summation is done over all individual backscattering cross sections (σ_i) over the area. Using this concept, we can derive another form of radar equation for radar clutter which shows that the power received from distributed clutter varies inversely as the range to the clutter raised to the *third* power.

In conclusion, point targets are an important source of echo for many radars. By making careful measurements of the return from point targets (range, azimuth, elevation, power, velocity, and other characteristics), much can be learned about the targets. Well-chosen point targets also make it possible to monitor the health and quantitative reliability of a particular radar system.

Chapter 5

Distributed Targets

When a radar is aimed at a meteorological target, there are many raindrops or cloud particles within the radar beam at the same time. Storms and clouds are usually so large that they completely fill the radar beam. The only place where this is not true is along the boundary of a storm where the radar beam will be moving from no echo to echo or vice versa. Similarly, near the top or bottom of a storm, the beam can be partially in and partially out of echo. Otherwise, when the beam is completely filled, the power returned to the radar will come from all of the individual targets being illuminated by the radar beam.

Let us consider the number of particles in the pulse volume for a moment. Continental clouds contain as many as 200 or more cloud droplets/cm^3. That amounts to $2 \cdot 10^8/\text{m}^3$. For a radar with a 1° antenna beamwidth, the beam will be 1 km in diameter at a range of 57 km. If the radar is using a 1-μs pulse length, the effective sample volume in space will be 150 m. The volume of the radar pulse is then illuminating more than $2 \cdot 10^{16}$ cloud droplets simultaneously.

The number of precipitation-sized particles is lower than this. Typical rain will have on the order of a few to a few hundred raindrops per cubic meter. Thus, there might be something like 10^9 to 10^{12} raindrops in a single radar sample volume. This is still a very large number of particles. The return from meteorological targets is the combination of billions of returns being averaged together.

Mathematically, we can express this quite simply by saying that the total backscattering cross-sectional area of a meteorological target is the sum of all of the individual backscattering cross-sectional areas, i.e.,

$$\sigma_t = \sum_{i=1}^{n} \sigma_i \qquad (5.1)$$

where the summation is carried out over all n particles in the sample volume.

Time to independence

One thing we need to be concerned about is how quickly we sample a volume of raindrops. If we send a pulse of radar energy into a storm and get an echo back and then send a second pulse into the storm immediately after the first, there would be little time for the raindrops to change position relative to each other or relative to the radar. If the pulses were sent nearly simultaneously, the returns measured by the radar would be virtually identical. If, on the other hand, we waited a reasonable length of time before sending a second pulse into the same point in space, the arrangement of particles being sampled by the radar might be totally different. Between these two limits is a region of interest and importance for radar.

Thus, when sampling raindrops or other hydrometeors with radar, we need to wait long enough to allow the particles to reshuffle enough so a truly different can be reached. Otherwise we are simply making multiple measurements of the same initial arrangement, and no new information has been gained. One reason we might want to do this is to get a good average of the true signal amplitude. Weather echoes are constantly changing. A single, instantaneous measurement might not be a good measure of the true signal strength. By averaging several samples together, we get a better measure of a storm's intensity.

The time it takes hydrometeors to rearrange themselves so the measurements are independent of one another is called the "time to independence" or the "decorrelation time." Mathematically, it is defined as the time it takes for a sample of targets to decorrelate to a value of 0.01 from perfect correlation. Since perfect correlation has a correlation coefficient of 1.0 (or –1.0 if there is an inverse correlation) and a correlation coefficient of 0.00 means there is no correlation at all, waiting for a signal to decorrelate to 0.01 means we have waited long enough that the newest sample is almost completely different than the original sample.

The decorrelation time of a sample depends upon three factors. One is the wavelength of the radar being used; a second is the hydrometeors themselves; and the third factor is the turbulence within the sample volume. When a short wavelength is used, particles do not have to travel as far to change position significantly relative to the radar, so decorrelation times are shorter for shorter wavelength radars.

The particle size distribution is a second factor. If all the particles are of the same size, they will all fall with the same terminal velocity and will fall together. This tends to make the sample decorrelate slowly. When the storm contains a wide variety of particle sizes, there will be many different sized particles falling in the same volume. This tends to make the sample decorrelate faster. The shortest decorrelation times occurs when hail and rain are in the same sample volume.

Studies of decorrelation times have found that $t_{0.01}$, time required for the autocorrelation function to fall to a value of 0.01, is approximately $t_{0.01} = 2\lambda$ to $t_{0.01} = 3\lambda$, where $t_{0.01}$ is given in milliseconds and λ is in centimeters. Measured decorrelation times have ranged from 3.5 ms to nearly 30 ms, depending upon the storm and radar.

If we want to sample as close together in time as possible but still have independent samples, we would want to sample at a rate given approximately by $10t_{0.1}$. This suggests we should sample at rates on the order of 3 to 30 times a second (3-30 Hz). Most radars sample at rates much higher than this. Modern Doppler radars often use pulse repetition frequencies (*PRF*'s) near 1000 Hz. At a *PRF* of 1000 Hz, the sampling time is much too close together to have truly independent samples, so it is necessary to average many consecutive pulses together in order to have the equivalent of just a few independent samples.

Without going into detail, there are other factors which can contribute to decreasing the time to independence of consecutive samples made with a radar. The factors include doing range averaging, moving the antenna in azimuth while collecting the data (this is almost always done anyway), wind shear within the sample volume, and turbulence.

Sample volume

As implied earlier, the radar samples a certain volume in space. This sample volume is given by

$$V = \pi \; \frac{r\,\theta}{2} \; \frac{r\,\phi}{2} \; \frac{h}{2} \qquad (5.2)$$

where θ and ϕ are the horizontal and vertical beamwidths, respectively, r is the distance to the sample volume from the radar, and h is the pulse length. While θ and ϕ are typically measured or quoted in degrees, they must be in radians in this and other equations that use them. V, r and h can be in any consistent set of units.

The pulse length h is the length in space corresponding to the duration τ of the transmitted pulse (be careful: the pulse duration of the radar is sometimes called pulse length). That is, $h = c\tau$, and c is the speed of light. In the equation for sample volume above, we used ($h/2$) because we are interested only in signals that return to the radar at precisely the same time. Since the front edge of a radar pulse starts τ seconds before the trailing edge, and since we want it to return back at the radar at the same instant as the trailing edge, it can only travel a short distance farther than the trailing edge. And since it must go out and back within τ seconds, the leading edge can travel only a distance $h/2$ before it starts its return trip; otherwise it would not arrive back at the radar simultaneously with echo from the trailing edge.

We want to know the total backscattering cross-sectional area of targets within the radar sample volume. A convenient way to do this is to determine the backscattering cross-sectional area of a *unit* volume and multiply this by the *total* sample volume. Thus,

$$\sigma_t = V \sum_{vol} \sigma_i \qquad (5.3)$$

Chapter 5

where $\sum_{vol} \sigma_i$ is the summation of the individual backscattering cross-sectional areas over a *unit volume.*

In Eq. 5.2 for sample volume we used the horizontal and vertical beamwidths θ and ϕ. This assumes that all of the energy in the radar's transmitted pulse is contained within the half-power beamwidths as used above. As was discussed earlier, however, real radar antennas do not have such nicely behaved beam patterns. Probert-Jones (1962) was the first to recognize this and derived a radar equation which correctly accounted for the power distribution within the mainlobe of antenna beams generated by the circular parabolic reflectors used with most meteorological radars. Using a Gaussian shape for beam pattern, Probert-Jones found the volume of a radar pulse volume to be the following:

$$V = \frac{\pi r^2 \theta \phi \, h}{16 \, ln(2)} \qquad (5.4)$$

where the additional factor of $2 \, ln(2)$ in the denominator accounts for the real beam shape better than the assumptions used in deriving Eq. 5.2 did. The term $ln(2)$ is the natural logarithm of 2 (i.e., logarithm to the base e).

Radar equation in terms of σ_i

Now, we can substitute our expressions for total backscattering cross-sectional area σ_t (Eq. 5.3) and sample volume (Eq. 5.4) into the equation for a point target derived before (Eq. 4.7) to get a radar equation for a beam-filling meteorological target. Doing this gives

$$P_r = \frac{P_t \, g^2 \, \lambda^2 \, \theta \, \phi \, h \sum \sigma_i}{1024 \, \ln(2) \, \pi^2 \, r^2} \qquad (5.5)$$

where all of the numerical terms have been combined.

Radar reflectivity η

Early in the history of the application of radar for meteorological targets, a parameter was defined that is related to the total backscattering cross-sectional area $\sum \sigma$. The parameter defined was named radar reflectivity and given the symbol η. Radar reflectivity was defined as follows:

$$\eta = \sum_{UnitVolume} \sigma_i \qquad (5.6)$$

where the summation is done over a unit volume of space. Since backscattering cross-sectional area has units of area (e.g., cm^2) and volume has units volume, radar reflectivity η typically has units of 1/cm or cm^{-1}. Radar reflectivity is an *intensive* parameter rather than an extensive parameter. $\sum \sigma_i$ is the total of all individual targets in the sample volume, but η is normalized to a unit volume. We'll return to radar reflectivity shortly, but first let's complete the derivation of the radar equation as it is commonly used by most meteorologists today.

Another aspect of the target size is related to the backscattering cross-sectional area. One of the complications was discussed earlier, that of the relative size of the target compared to the wavelength of the radar. If the particles are small compared to the wavelength, the Rayleigh approximation applies. If they are large compared to the wavelength, the targets will be in the optical region. And if they are intermediate, they will be Mie scatterers. For most meteorological radars (i.e., wavelengths of 3 cm and larger, almost all raindrops can be considered small compared to the wavelength, so the Rayleigh approximation applies. Recalling Eq. 4.9

$$\sigma_i = \frac{\pi^5 |K|^2 D_i^6}{\lambda^4} \qquad (5.7)$$

where σ_i is the backscattering cross-sectional area of the ith sphere; λ is the wavelength; D is diameter; $|K|$ is the magnitude of the parameter related to the complex index of refraction. It is given by

$$K = \frac{m^2 - 1}{m^2 + 1} \qquad (5.8)$$

The complex index of refraction m is given by

$$m = n + ik; \qquad (5.9)$$

n is the index of refraction of the sphere, $i = \sqrt{-1}$, and k is the absorption coefficient of the sphere.

The value of $|K|^2$ depends upon the material, the temperature and the wavelength of the radar. The temperature and wavelength dependencies are not very large but might need to be accounted for in precise work.[6] The dependence upon the kind of material, however, is significant. For the most commonly used

---

[6] Battan (1973) gives two tables that list the appropriate values of $|K|^2$ for water and ice at various temperatures and for wavelengths of 0.62, 1.24, 3.21 and 10 cm.

radar frequencies and over reasonable temperatures, $|K|^2$ for water is usually taken at a value of 0.93. For ice, $|K|^2$ = 0.197. These two values differ by a factor of 5 or about 7 dB. It is fairly common practice to assume that every part of a storm being scanned by radar is made up entirely of water, so the value of $|K|^2$ for water is used to calculate radar return. When we know for sure that we are making measurements from a region of a storm which is ice, then it is necessary to use the value of $|K|^2$ for ice instead; otherwise our reflectivities will be off by 7 dB.

Radar equation in terms of D^6 and z

If we substitute Eq. 5.7 for σ_i into Eq. 5.5, we get the following

$$p_r = \frac{\pi^3 p_t g^2 \theta \phi h |K|^2 \sum D_i^6}{1024 \ln(2) \lambda^2 r^2} \qquad (5.10)$$

This equation is perfectly fine for calculating the power received from a sample of raindrops, providing we know the diameters of all of the raindrops in a unit volume. Obviously, this is not going to happen most of the time. So, to get around this problem, we define one final parameter called the radar reflectivity factor

$$z = \sum_{vol} D^6 \qquad (5.11)$$

Again, the summation is carried out over a unit volume, not over the total sample volume of the radar. If we make this substitution, we have a useful version of the radar equation:

$$p_r = \frac{\pi^3 \, p_t \, g^2 \theta \, \phi \, h \, |K|^2 \, z}{1024 \, \ln(2) \, \lambda^2 \, r^2} \qquad (5.12)$$

This equation is quite general; it can be applied to any radar, provided the particles meet the Rayleigh assumption.

Now let's return to radar reflectivity η and compare it to radar reflectivity factor z. If we examine Eqs. 5.6, 5.7 and 5.11, we can see that η and z are related through

$$\eta = \frac{\pi^5 |K|^2 z}{\lambda^4} \qquad (5.13)$$

Notice that the dimension for z is L^3 (where L is "length") while that for λ^4 is L^4, so η has dimensions of L^{-1}.

Why do we need two parameters to describe meteorological targets? Wouldn't either one of these serve the purpose just fine? The answer to the second question is "yes". In fact, both are still used for different purposes. Radar reflectivity was defined first and was used in many early meteorological studies. But it has a serious disadvantage. Radar reflectivity η from a storm depends upon the wavelength of the radar making the measurements. Radar reflectivity factor z does not have this same restriction. As a result, the radar reflectivity factor z of a storm is independent of the radar; it is truly a property of the storm. This should be obvious from the definition of z given by Eq. 5.11. There it is clear that z only depends upon the number and sizes of the raindrops in the storm.

One of the disadvantages to having two reflectivity parameters is that the simplest name got used first. Nowadays, radar reflectivity factor z is used most frequently, but the simple name "radar reflectivity" was already used. Now, three or four decades later, many of us conveniently solve this problem by ignoring the original radar reflectivity η and refer to z as radar reflectivity. And we often even leave off the word "radar", since almost all of the time we know that the term we call "reflectivity" is really "radar reflectivity factor." We're lazy, I suppose, but as long as we know what we are talking about, it should not create any serious problems.

Now let's continue with our derivation of a beam-filling meteorological radar equation. There is still one term missing from Eq. 5.12. We have not accounted for attenuation. As will be discussed in Chapter 8, the power returned to a radar can be reduced by a number of factors. This loss of power in traveling through a medium (e.g., the atmosphere, cloud, rain, snow, hail, or even through a waveguide or radome) is called attenuation, and it *always* reduces the power received by the radar. As will be seen later, the magnitude of the attenuation from specific sources can be quantified fairly easily some times. We will postpone the quantitative aspects of attenuation until later. However, to complete our discussion of the radar equation, we can add a final term to account for the total loss due to all attenuation. This term will be given as l. The attenuation term l is always between zero and one, and usually closer to 1 than 0 (otherwise we would not be able to detect many of the targets we are interested in). So, our final radar equation can be written

$$p_r = \frac{\pi^3 \, p_t \, g^2 \theta \, \phi \, h \, |K|^2 \, l \, z}{1024 \, \ln(2) \, \lambda^2 \, r^2} \tag{5.14}$$

Since the attenuation is often unknown for a given situation or because we choose to ignore it, the attenuation term is often omitted unless attenuation is specifically being considered in the calculations. In the remainder of this text, I will not be included except when it is being discussed.

For a given radar operating normally, we can simplify the radar equation considerably. All of the parameters associated with a specific radar can be grouped together as a constant. This includes p_t, g, θ, ϕ, h, and λ. Further, we can combine the numerical constants (π, 1024, and $\ln(2)$). If we do all this, we can write the radar equation as

$$p_r = \frac{c_1 \, |K|^2 \, z}{r^2} \qquad (5.15)$$

Then, if we specify that we are primarily interested in looking at liquid hydrometeors rather than ice, we can substitute in an appropriate value for $|K|^2$. Doing this gives the radar equation as

$$p_r = \frac{c_2 \, z}{r^2} \qquad (5.16)$$

where this constant c_2 is different from the one above.

Equation 5.16 is the working equation for beam-filling meteorological targets. It says that the power received by a given radar is proportional to the radar reflectivity factor of the storm and inversely proportional to range squared. The stronger the storm, the stronger its reflectivity will be and the higher the power received by the radar.

The range variation is also significant. Students of physics will recognize this as the familiar "inverse square law" which applies to radar waves as well as to light. As a storm gets farther from the radar, the power returned to the radar decreases even more rapidly. Two storms of equal reflectivity will give equal powers back only if they are at the same range.

Since what we are primarily interested in is radar reflectivity factor z, we can rearrange this equation and get an equation of the form

$$z = c_3 \, p_r \, r^2 \qquad (5.17)$$

This equation contains the constant c_3 which can be called the "radar constant". c_3 has units of $mm^6/m^3 \, mW^{-1} \, km^{-2}$.

Let us consider radar reflectivity factor z for awhile. Reflectivity is a meteorological parameter that is determined by the number and size of the particles present in a sample volume. It can range from very small values in fog (perhaps $0.001 \, mm^6/m^3$) to very large values in very large hail. The highest radar reflectivity factor I have ever seen was $36,000,000 \, mm^6/m^3$ in a hailstorm in Montana (1981) when hail as large as softballs was falling. Because of the tremendous range of values that z can have, it is convenient to compress this over a smaller range of numbers. One way to do this is to use logarithmic values instead of linear values. Thus, we can define the logarithmic radar reflectivity factor Z as follows

$$Z = 10 \log_{10}\left(\frac{z}{1mm^6/m^3} \right) \qquad (5.18)$$

where Z is the logarithmic radar reflectivity factor measured in units of dBZ (i.e., decibels relative to a reflectivity of 1 mm^6/m^3), and z is the linear radar reflectivity factor in mm^6/m^3).

Using the logarithmic reflectivity has the advantage of compressing the range of values given above for extremes to much more convenient numbers. The examples given become –30 dBZ for fog and +76.5 dBZ for large hail on a logarithmic scale. These are much easier to use (and, I believe, to comprehend) than the much smaller or much larger values of the corresponding linear values.

Since radar reflectivity factor is most commonly measured in logarithmic units, we can convert Eq. 5.17 to logarithmic form, giving

$$Z = C_3 + P_r + 20 \, log_{10}(r). \tag{5.19}$$

where radar reflectivity factor Z will be measured in dBZ, received power P_r is measured in dBm, r is in kilometers, and the constant $C_3 = 10 \bullet log_{10}(c_3)$. Appendix D gives the logarithmic radar constant for a number of different meteorological radars. And Appendix A gives a more complete discussion of logarithmic units, powers in dBm, and reflectivities in dBZ.

Effective radar reflectivity factor z_e

As mentioned already, we frequently are lazy and simply call this parameter "reflectivity." As long as this does not cause any confusion, it is usually acceptable. But in introducing Eq. 5.7 which relates z to backscattering cross-sectional area σ, we specified that the targets must be spheres and that these spheres must be small compared to the wavelength. When we scan a storm with a radar, we cannot always be sure that this is true. At other times we know or strongly suspect that it is not true. To accommodate these possibilities, we define a slightly different term which we call the "equivalent" (or "effective") radar reflectivity factor and give it the symbol z_e or Z_e. Anytime we measure reflectivity with a radar, we might consider using equivalent radar reflectivity factor instead of z or Z.

Z from drop-size spectra

Throughout this discussion we have tacitly assumed that we are going to measure reflectivity with a radar. There is another way to calculate z, however, and this is to use a drop size distribution. That is, we can use Eq. 5.11 to calculate radar reflectivity factor.

A drop-size distribution tells us how many drops of each size are present in a sample volume. We can make size distribution measurements a number of different ways. One of the simplest is to take a piece of water-absorbing paper, cover it with a dusting of a chemical which changes color when it dissolves in water, and then expose it to rain for a measured period of time. Raindrops hitting the paper will form colored spots which can then be measured. By calibrating the paper used, we can convert the spot diameter into the diameter of the drop which formed the spot. By measuring the time of the exposure and knowing how fast raindrops of a particular size fall, we can calculate the sample volume for each given diameter.

In making drop-size distributions it is impossible to measure individual drops to infinitely fine precision. Instead, we usually quantize our measurements into small diameter intervals. For example, since raindrops range in diameter from a fraction of a millimeter up to about 5 mm, we can measure diameters to the nearest 0.2 mm or 0.5 mm and get a useful set of measurements.

Then, given a drop-size distribution from a sample of rain we can calculate the radar reflectivity factor z by using

$$z = \Sigma N_i D_i^6 \tag{5.20}$$

In this case we have included the term N_i which is the number of drops of diameter D_i to $D_i + \delta D$ where δD is the diameter interval used in making the measurements.

Appendix E describes how to calibrate and use the filter-paper technique for making drop-size distribution measurements. It also includes a template that can be used to measure raindrops providing the same kind of paper is used as was used to determine the calibration.

NO, NANCY. THAT'S DECIBEL, NOT DECIBULL.

Chapter 6

Doppler Velocity Measurements

Most of what has been discussed until now in this text has dealt with measurements of echo power from a radar. These are used to determine the backscattering cross-sectional area of point targets and the radar reflectivities and rain rates from meteorological targets (see Chapter 8). Of course, conventional, reflectivity-only radars also provide very useful information on the positions of storms, their movement, development and other properties. Many radars, however, now provide direct measurements of the speed of movement of by using the Doppler effect mentioned earlier.

Christian Doppler discovered that a moving object will shift the frequency of sound in proportion to the speed of movement. The classic example is that of the train whistle approaching a stationary observer. If the train blows its whistle while approaching and continues blowing it as it passes and goes off in the other direction, anyone listening to the sound will hear the pitch of the whistle decrease when the train passes.

Exactly the same thing happens with electromagnetic radiation as happens with sound. In the case of sound, however, the frequency shift was usually noticed by having a moving source and a stationary listener. In the case of radar, the usual situation is to have a stationary radar observing moving targets. Each target that is moving will shift the frequency of the radar signal an amount which depends upon its speed.

Consider a single target at distance r from a radar. The total distance a radar pulse will have to travel to detect this target is $2r$ since the wave has to go out to the target and back to the radar.

$$total\ distance = 2r$$

This distance can also be measured in terms of the number of wavelengths from the radar to the target.

$$distance\ in\ wavelengths = 2r/\lambda$$

where λ is the wavelength of the radar signal. We can also measure this distance in radians by using the fact that 1 wavelength $= 2\pi$ radians. So,

$$distance\ in\ radians = (2r/\lambda)2\pi$$

The phase of an electromagnetic wave is essentially the fraction of a full wavelength a particular point is from some reference point measured in radians or degrees (see Fig. 6.1). The reference point of a wave is usually the point on a sine wave where the cosine is one and sine is zero. If our point of interest is at the very beginning of the sine wave, its phase is zero. If it is a quarter of the way from the start of the wave toward the next wave, its phase is $2\pi/4 = \pi/2$ radians (90°). Phase shifts can be either positive or negative but are always less than 2π radians (360°). In fact, if the phase shift is more than π radians (180°) we usually consider the phase difference to be the difference in angle between our point of interest and the nearest reference point. That way, our phase shift will never be more than $\pm\pi$ radians or $\pm180°$. The second (dashed) curve on Fig. 6.1 has a phase shift of 30° from the solid curve.

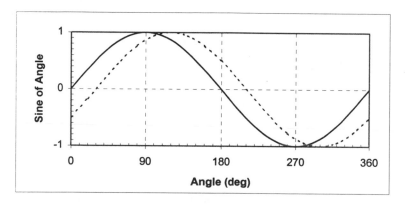

Figure 6.1 Sine wave (solid curve) and a second signal 30° out of phase with the first wave (dashed curve).

If a radar signal is transmitted with an initial phase of ϕ_0, then the phase of the returned signal will be

$$\phi = \phi_o + \frac{4\pi r}{\lambda}. \qquad (6.1)$$

The change of phase with time from one pulse to the next is given by

$$\frac{d\phi}{dt} = \frac{4\pi}{\lambda}\frac{dr}{dt} \qquad (6.2)$$

Chapter 6

where $d(\ )/dt$ is the time derivative or time rate of change of the parameter. The velocity of an object is given by

$$V = \frac{dr}{dt}. \tag{6.3}$$

Angular frequency Ω is the time rate of change of angle (or phase) and is defined by

$$\Omega = \frac{d\phi}{dt}$$

$$= 2\,\pi f, \tag{6.4}$$

where f is the frequency shift in cycles per second (hertz). Thus, by combining Eqs. 6.2, 6.3, and 6.4, we get the frequency shift caused by a moving target;

$$f = \frac{2V}{\lambda}. \tag{6.5}$$

So a given phase shift $(\delta\theta)$ in a given interval of time (δt) becomes a frequency shift which the radar can measure.

This, then, is the frequency shift caused by a target moving relative to a radar. Notice that it is linearly proportional to velocity and inversely proportional to wavelength. For a given radar, wavelength is a constant, so the frequency shift is dependent only upon velocity of the target.

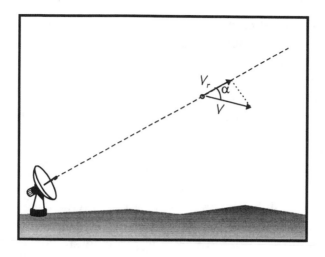

Figure 6.1b Geometric relationship of a target located on the center of the antenna beam pattern and moving with velocity V at an angle α relative to the direction the antenna is pointing. The radar detects the radial component of velocity V_r.

If the target is not moving directly toward or away from the radar, we can easily correct for this by determining the *radial component* of motion using

$$f = \frac{2V\cos(\alpha)}{\lambda}.$$
(6.6)

where angle α is shown in Fig. 6.1b.

Block diagram of Doppler radar

Now let's examine how a Doppler radar measures the speed of a target. Figure 6.2 shows the block diagram of a pulsed-Doppler radar. It has all of the components as the simple radar given earlier (Fig. 2.1), but it also has some additional components. The ability of a Doppler radar to detect slight phase shifts depends critically upon the system maintaining a constant transmitter frequency and phase relationship from one pulse to the next. Many Doppler radars use what are known as coherent transmitters (generally using klystron transmitting tubes). These radars transmit exactly the same frequency and initial phase from one pulse to the next. Other Doppler radars, using magnetron transmitting tubes, do not maintain the same frequency and phase stability but have components which sample and remember the phase and frequency of each pulse so that it can be compared with the received signal. Both kinds of transmitters give excellent velocity measurements.

In order to determine the frequency shift from one pulse to the next, Doppler radars contain a device called a stable local oscillator (STALO) which maintains a very stable frequency from one pulse to the next. The signal from the STALO is mixed with the frequency from the transmitter in a locking mixer. This signal is sent through a coherent oscillator which amplifies this signal while maintaining the phase relationship with the initially transmitted signal. The signal from the STALO is also mixed with the received signal in the receiver/mixer. This signal is amplified in the intermediate frequency (IF) amplifier. The received signal and the sample of the transmitted signals are sent to a phase detector which compares the phases of the two signals and determines how much the received signal has been shifted relative to the transmitted signal. This is processed and displayed and/or recorded by additional components in the system.

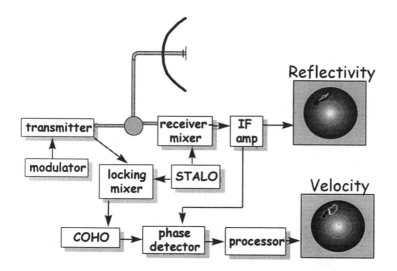

Figure 6.2 Block diagram of a simple Doppler radar. IF amp = intermediate frequency amplifier; STALO = stable local oscillator; COHO = coherent oscillator. The top "display" shows radar reflectivity factor while the bottom "display" show Doppler radial velocity.

Maximum unambiguous velocity

There are limitations in the velocities and ranges that a radar can resolve unambiguously. Let us consider velocity ambiguities first. When a target is not moving toward or away from a radar, it will have zero *radial* velocity. This does not necessarily mean that the target is stationary. It simply means that the target is remaining at a constant distance from the radar. It could be moving quite rapidly, in fact, but any movement it has must be perpendicular to the radar's beam. Since the only velocity a Doppler radar can detect using phase-shift principles is the radial velocity, we usually omit the qualifier "radial" and simply talk about the "velocity" (see Fig. 6.1 again). While this is convenient, be careful to recognize that a Doppler radar detects only radial velocities.

If the velocity of a target relative to a radar is zero, there will be zero phase shift in the frequency of the received signal relative to the frequency of the transmitted signal. If the target is moving slightly away or toward the radar, there will be a slight phase shift. As the speed of the target increases, the phase shift will also increase, producing an increasing Doppler frequency shift. There is a limit, however, of how large a phase shift a radar can detect. For example, if a target were moving away from a radar just fast enough that it traveled 1/2 a wavelength between two consecutive radar pulses, it would produce a phase shift of π radians. If it were moving toward the radar at the same velocity, it would also produce exactly the same shift of π radians, and we could not tell the difference in their velocities. As another example, if a target were moving so fast that it traveled exactly a whole wavelength between two consecutive pulses, the radar would detect zero phase shift and think that the target was stationary.

The maximum velocity a Doppler radar can detect correctly or unambiguously is given by the velocity which produces a phase shift of π radians. This is also called the Nyquist frequency or Nyquist velocity[7], depending upon whether we are referring to the maximum unambiguous frequency or velocity, respectively. Mathematically, we can express this as

$$V_{max} = \frac{f_{max}\,\lambda}{2},\qquad (6.7)$$

---

[7] Harry Nyquist received his BA and BSEE in 1914 and his MA in 1915, all from the University of North Dakota.

where the maximum frequency f_{max} is given by

$$f_{max} = \frac{PRF}{2} \tag{6.8}$$

and PRF is the pulse repetition frequency of the radar. Thus, the maximum unambiguous velocity detectable by a Doppler radar is

$$V_{max} = \frac{PRF\ \lambda}{4} \tag{6.9}$$

This is an important result. It says that if we want to be able to detect high velocities, we must use long wavelengths, large PRF's or both.

Maximum unambiguous range

Now, what about the limitations on range? We know that electromagnetic radiation travels at the speed of light. The time it takes for a signal to go out to and back from a target is $t = 2r/c$, where c is the speed of light, r is range and t is time. The "2" accounts for the distance out and back from the target. If a radar transmitted a single pulse and waited forever for a returning echo, there would be no limit to the distance at which the correct range to a target could be determined. In the real world, we do not wait very long before sending out a second pulse.

There are a number of reasons for this. One is that we cannot detect targets at extremely long ranges or we are not interested in them. Meteorological targets typically exist only 10 to 15 km above the earth's surface. Even though the radar waves bend downward somewhat in their travel through the atmosphere, the earth's surface curves away even faster, so the radar beam usually gets so high above the earth's surface that storms are not detectable beyond 400 to 500 km from a ground-based radar.

Another reason we are not interested in distant targets is that the inverse square law decreases the power received from a meteorological target according to $1/r^2$. If a target is too far away, the power received from it will be so weak that the radar will be unable to detect it. For these and other reasons, radars are designed to send out subsequent pulses of energy at fairly frequent intervals.

A radar transmits many pulses each second. The rate is given by the PRF. The time T between pulses is thus

$$T = \frac{1}{PRF}.$$ (6.10)

Now, given T, we can determine the maximum range a radar signal can travel and return before the next pulse is sent out. This is simply

$$r_{max} = \frac{cT}{2}$$

$$r_{max} = \frac{c}{2\,PRF}.$$ (6.11)

The Doppler dilemma

The combination of maximum unambiguous velocity and maximum unambiguous range form two constraints which must be considered in choosing the PRF for use with a Doppler radar. Notice that non-

Doppler radars are only constrained by the maximum unambiguous range; since they cannot measure velocity, the velocity constraint does not apply.

Unfortunately, PRF appears in both Eqs. 6.9 and 6.11, but in the denominator of one and the numerator of the other. This forms what has been called the "Doppler dilemma." By solving both equations for PRF and equating them, we find that

$$V_{max} \, r_{max} \; = \; \frac{c \, \lambda}{8} \qquad (6.12)$$

If we want to have a large V_{max}, we must have a small r_{max} since the right side of the equation is a constant for a given radar. Conversely, if we want to detect echoes at long ranges, we can only detect small velocities.

Figure 6.3 (based on Gossard and Strauch, 1983) shows the Doppler dilemma graphically. Note that the ordinate (Y-axis) on this figure gives the maximum velocity interval corresponding to the Nyquist frequency. Normally we divide this interval in half with the maximum unambiguous velocity being divided into plus *and* minus half of the V_{max} interval. For example, from the figure we can see that for an S-band radar, if the PRF is 1000 Hz, the maximum unambiguous range is 150 km while V_{max} is ±25 m/s. For an X-band radar using the same PRF, r_{max} is still 150 km, but V_{max} is now only ±8 m/s. For meteorological situations, we may want to measure velocities as large as ±50 m/s out to ranges beyond 200 km, so neither of the limits calculated above is completely adequate. The S-band system comes much closer to being useful than the X-band system, however. And C-band will be intermediate to these two.

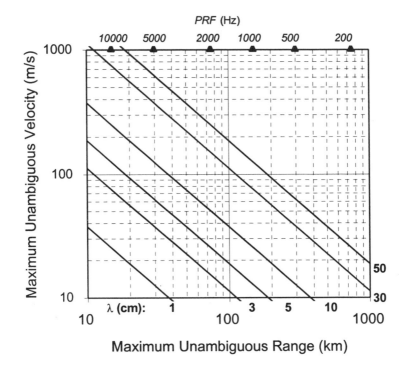

Figure 6.3 Summary of conditions for range and velocity folding (i.e., the Doppler dilemma). The numbers near the bottom are wavelength. Those near the top are pulse repetition frequency. Notice that r_{max} depends only on PRF and is independent of wavelength.

One partial solution to the Doppler dilemma is in our choice of wavelength. We can increase both V_{max} and r_{max} by using a longer wavelength radar. Unfortunately, longer wavelength radars are more expensive and bigger, and they don't detect weather targets as well as shorter wavelength radars, so using a longer wavelength is not necessarily a solution to the problem. The result is that most Doppler weather radars usually suffer significant range or velocity ambiguities or both.

Since range ambiguities (also called aliasing or folding) are so common with modern Doppler radars, let us examine the causes of this in a little more detail. Range aliasing occurs because we don't wait long enough between transmitted pulses. Instead, we transmit pulses close together (mostly to make the Doppler side of the radar work better), not giving one pulse enough time to cover the distance between the radar and some storms before the radar sends out the next pulse of energy. Consequently, a storm can be detected beyond r_{max} (see Fig. 6.4). That is, a first pulse of energy goes beyond r_{max} and detects a storm. The radar does not know the signal is from beyond r_{max}, however, and displays it at a distance of $(r - r_{max})$.

Chapter 6

Echoes which are displayed in the wrong range interval are called multitrip or second-trip echoes. If the PRF is high enough and distant echoes tall enough and strong enough, sometimes third or even fourth trip echoes can be detected.

Recognizing range-aliased echoes

How are second trip echoes recognized on a radar? There are a number of ways multitrip echoes can be recognized. One of the easiest is to simply look outside and see what is going on in the real world. If the radar shows a nearby storm in a particular direction but there is nothing outside, it is probably a multitrip echo.

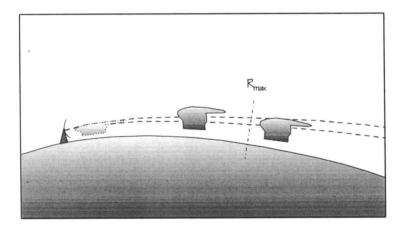

Figure 6.4a Illustration of how a storm beyond r_{max} can be displayed at the wrong range. Two real echoes exist. The first is less than r_{max} away and is displayed at the correct range. The second is beyond r_{max}; it is displayed at a range of $(r - r_{max})$. The faint, dashed storm near the radar is where the radar would display the distant storm.

A second way to recognize multitrip echoes is by their shapes (see Fig. 6.4a and 6.4b). Real storms are usually somewhat circular, elliptical, or irregular. Storms certainly should not know where the radar is located. Anytime a narrow, wedge-like echo is detected which points toward the radar, second-trip echoes should be suspected.

Another clue to the existence of multitrip echoes is height (see Fig. 6a). Real echoes, especially from convective storms, usually extend up into the

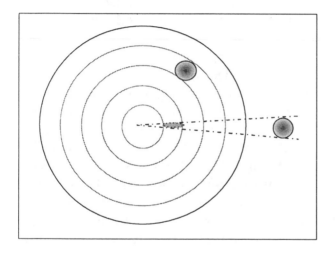

Figure 6.4b Simulated PPI display showing a real echo located to the northeast. To the east is an echo beyond r_{max}. It is displayed at a distance of $r - r_{max}$ from the radar. It also has a very narrow shape. Both the range and azimuthal extents are exactly the same in the real and the aliased positions, but its aliased azimuthal distance is much narrower. Also, its reflectivity will be weaker because of the $1/r^2$ dependence on received power in the radar equation.

atmosphere several kilometers. Thunderstorms are frequently 8 to 15 km in height. If a convective-like echo appears on the radar display but it has an indicated height which is much less than normal, it may be a second trip echo. For example, a real thunderstorm which is 10 km tall at a range of 200 km would be detectable at an elevation angle of about 2.2° (see Fig. 3.2). If it is a second trip on a radar with a PRF of 1000 Hz, it would show up at 200 km - 150 km = 50 km. If the echo from this storm disappears at 2.2°, its indicated height would only be 2 km. This is a ridiculously small height for a strong storm, so you should expect range aliasing.

Finally, second trip echoes can sometimes be recognized by their reflectivities. The power received from a storm decreases according to $1/r^2$. If our storm being displayed at 50 km were real, it would have a certain reflectivity. If it is really at 200 km, however, the power returned from it would be $(200/50)^2$ less than if it were at 50 km. So the returned signal would be 16 times less. On a decibel scale this would be 12 dB less than if it were at its indicated range. Unfortunately, since we do not know the true reflectivity of a storm without the radar giving it to us, we cannot be sure that a weak echo is simply a weak storm and not a second-trip storm. Nevertheless, low reflectivity combined with shape and height information can help differentiate real from multitrip echoes.

There is one guaranteed-or-double-your-money-back way to unambiguously determine if echoes are range aliased or not: *Change the PRF!*. If we change the PRF and watch the positions of echoes, all correct echoes will not change their range whereas range-aliased echoes will shift in or out in range, depending upon whether the PRF is increased or decreased.

Alternatively, we can avoid range aliased echoes by using a PRF so low that r_{max} is so large that range aliasing cannot take place.

Some examples of second-trip echoes are shown in Color Figs. 6, 7, 9 and 10. In Color Fig. 6 there is some real echo toward the west to northwest near 60 km. The second-trip echoes are visible mostly on the reflectivity display and are weak echoes extending along radials from near the maximum range in toward the nearby ground clutter. Further evidence that the suspicious echoes are really second-trip echoes is that they do not show much at all in the velocity data. The signal processor on the UND radar rejects velocity data that do not meet certain quality checks (i.e., signal strength above some threshold and velocity variance smaller than some upper limit plus another check or two); second-trip echoes frequently fail these tests and are not displayed; because the reflectivity data are not subject to the same battery of tests and because the reflectivity data come from a different receiver, they are not eliminated on the reflectivity display.

In Color Fig. 7 there are some weak (green) echoes in the reflectivity data pointing toward the radar from about 10 to 40 km range at an azimuth of about 290° to 310°. Unfortunately - and this is frequently the case - the second-trip echo occurs in the same place as real echo of importance. In this case, there is a weak gust front in this same location. Again, these second-trip echoes show best on the reflectivity display and barely at all on the velocity display.

Color Fig. 9 also has some suspicious echoes which appear to be second trippers toward the very south end of the figure. In this region there are several small wedge-like echoes pointing toward the radar. Their reflectivities are not particularly strong, generally not exceeding 25 dBZ. In this case, however, the suspicious echoes do have velocities associated with them. The reason for this is that the velocity quality tests have been turned off (as indicated by the "SQI" value of 0.0 in the radar housekeeping data; see the comments about the interpretation of the housekeeping data in the section on Color Figures later in this text). As a result, range aliased velocities are not filtered out of the velocity display. Their velocities do seem to fit with those of the major echo to the west of the radar reasonably well, so it is not absolutely clear that they really are second-trip echoes from this single PPI. However, there is a long line of very strong reflectivities approaching the radar from the west. When strong lines of echoes exist, they are often very long. It is likely that this line of echo seen on the limited area of the radar display has more echo on the south end that is not shown - except as second-trip echoes!

Further evidence that second trip echoes are a possibility in Color Fig. 9 is the narrow band of bright red velocity near 10 km range and 240° azimuth. This is a case of a velocity not fitting the surrounding area. Color Fig. 10 shows a very shallow layer of reflectivity near the surface that is not connected to the main storm echo. This echo, which is only a kilometer or so tall, has reflectivities near 40 dBZ, yet it does not appear to be ground clutter (because it seems we are seeing some weaker echo below it; also because it appears smooth rather than rapidly changing as most ground clutter does). This supports the suspicion that this (and possibly others) are second-trip echoes and not correctly displayed by the radar.

Velocity aliased echoes

Now let's return to velocity aliasing. Recall that velocity aliasing occurs when the phase shift is more than ±180° or ±π radians from the transmitted phase at a given range. But how does this appear on a radar display? And how does the speed of a target produce this much phase shift? To begin with, let's consider a simple case where a Doppler radar is detecting an echo in the presence of a uniform wind, that is, the wind is the same speed and direction at every point the radar looks. When this occurs, the radial velocity detected by the radar depends upon both the direction the wind is blowing and the azimuth the radar is pointing. Equation 6.6 applies, and it suggests that the radial velocity detected is a function of the cosine of the angle between the radar beam and the wind direction.

Most Doppler weather radars display fields of radial velocity on color displays, usually showing the velocities on a scale going from the maximum unambiguous negative velocity to the maximum unambiguous positive velocity as a series of colors. And usually the color bar that is the key to determining velocities on the display is shown as a horizontal or vertical bar (see the color figures for both the UND radar and the NEXRAD radar).

When an echo contains a region that is faster (either away or toward the radar) than V_{max}, it will usually be evident by having fast approaching and fast receding velocities next to each other. But the linear color bar used to describe this is not always helpful in explaining *why* the aliasing is occurring. One way to maybe make this more obvious is to change the color bars into a circular format. By doing this it becomes a little more obvious that the velocity scale is really a continuum which has an artificial boundary between the maximum receding and maximum approaching velocities.

Consider Figure 6.5 which illustrates such a Doppler "speedometer". It shows the radial velocity detected for a radar when the environmental wind is from the west at various speeds. In the first case (top panel and top speedometer), the wind is perhaps 5 m/s. The radial velocity detected by the radar varies sinusoidally with azimuth. The speedometer is showing the speed that would be detected looking toward the east. In this direction the radar would be showing the maximum receding velocity of about 5 m/s. In the middle panel, the speed is faster and the speedometer is showing a receding velocity near 10 m/s.

The bottom panel shows what happens when the environmental wind exceeds the maximum unambiguous velocity of the radar. Now the approaching speed is on the order of 17 m/s or so. But the speedometer would now show a speed on the order of -13 m/s, i.e., it now shows a fast *approaching* velocity. The correct velocities will be displayed from north through ENE and from ESE to WSW and from WNW to north again. Where the sine wave exceeds the shaded area on the right panels, it will be forced or aliased into the gray region, but will have a velocity of the opposite sign.

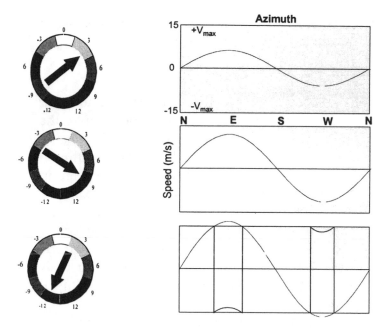

Figure 6.5. The "Doppler speedometer". The abscissa is the direction the antenna is pointing and the ordinate is the speed detected. In all three cases, the radar is detecting a uniform wind blowing from the west (made detectable, for example, by the presence of insects). For the case at top right, the velocity is about a third of V_{max}. For the middle case, the velocity is about two-thirds of V_{max}. For the bottom case, the speed is just over V_{max} and is aliased when the radar looks both upwind and downwind.

Notice that the color displayed by the radar for a particular echo is a function of the azimuth where the radar is pointing. If the radar is pointing perpendicular to the movement of the echo, the velocity displayed will be zero. This occurs on Fig. 6.5 where the sine wave crosses at zero velocity, i.e., north and south.

Recognizing velocity aliasing

How do velocity-aliased echoes appear on a radar display? The answer to this depends upon where the aliasing takes place. If a large region of echo is being detected by a Doppler radar and a region within it exceeds V_{max}, then there will be an abrupt change in velocities surrounding the aliased region. For example, if the storm is moving away and part of it is moving away faster than V_{max}, then strong receding velocities would surround a region with apparently strong approaching velocities. Such a discontinuity is usually quite visible, and it is obvious that velocity folding is taking place.

If the storm causing range folding is completely isolated such that there is no surrounding echo, the velocities from the storm may appear entirely correct even though they have been folded into the wrong velocities. This would make recognizing velocity-folded data much more difficult. Fortunately, such isolated situations are not very common, so this is usually not a major problem. There are almost always several echoes on a display at the same time (perhaps even more so when velocities are so strong as to be folded), so velocities of nearby echoes are often useful to indicate whether folding is taking place or not.

A more difficult situation, however, occurs when C- or X-band radars are measuring storm velocities. For these radars V_{max} can be moderately small. Thus, it is possible to have velocities which are not just folded once but are folded twice or more. This can make it extremely difficult to tell what the true velocities are from a quick visual inspection of the radar display.

Velocity aliasing is a very common feature with the UND radar. Except for Color Figures 1, 2, 6, and 12, *all* of the velocity figures from the UND radar have at least some velocity aliasing in them. In the 25 May 1989 example shown in Color Fig. 10, the velocities are so fast toward the radar that the velocities are wrapped around the color bar *twice!* This occurs, for example, in the region above 20 km range. Near the surface at 20 km range, velocity is quite strong (aliased from the reds into the blue) at about 15.5 m/s away from the radar. These receding velocities decrease to zero at about 2.5 km altitude above the radar and then become greater and greater toward the radar. At about 4+ km the blue velocities abruptly change to reds and continue in the same direction through zero and into the blues again; at about 18 km they again abruptly change into reds and peak out in the yellows. Above this they start to decrease back through the reds and blues. At the point where the velocities are strongest, they are approximately 55 m/s.[8]

A few (belated?) words are in order about how reflectivities and velocities are presented on modern radar displays. Most modern radars currently use color displays to show their data. We have already seen several examples of these. Reflectivities typically are shown using a color scale which has blues or greens for weak echoes and oranges, yellows and reds for stronger reflectivities. Depending upon the color resolution of the system, reflectivities might be displayed in as few as

---

[8] This value was obtained by calculating V_{max} using the PRF of 1100 Hz given on the color display; this gives $V_{max} = 14.9$ m/s. Then, since we went from zero to V_{max} and then all the way across the color bar to V_{max} again, we covered a total velocity range of $3 \cdot 14.9$ m/s = 44.6 m/s. In addition to that, the highest velocity reached was about 3 color bars beyond the left side, or $3 \cdot 3.3$ m/s ~10 m/s faster. Adding all these together gives 54.5 m/s. This was indeed a fast moving part of the storm!

three or four colors to as many as several hundred. A dozen to 16 colors is quite common. With storm intensity we know that as we go from outside the storm to inside the storm, the reflectivities must increase (from $0 \text{ mm}^6/\text{m}^3$ to z_{max}). Thus, we should always expect to see at least a narrow band of low reflectivities surrounding the stronger reflectivities.

Velocity displays usually show colors going away from the radar as browns, yellows and reds and those approaching as greens and blues. This convention was chosen to correspond to the "red shift" of visible light in our expanding universe. Since an entire storm may be going away or coming towards the radar, we cannot know ahead of time whether it should appear as all greens, all reds or some combination of the two. But we can usually expect to see a gradual transition from one color to the next. In principle, we should always see a gradual transition from one speed to another. This transition may take place over a very short distance, however, so it may appear fairly abrupt.

This normal progression is useful in looking for folded velocities. If the echo shows a region of near-maximum approaching velocity immediately adjacent to an area of near-maximum receding velocity, folding is almost certainly the cause.

Other reasons for velocity discontinuities

Discontinuities in the displayed velocity field, however, may come from at least three other sources other than range folding. If second-trip echoes are being detected and if they happen to overlay first trip echoes, the velocities in the second trip echo will probably not match those of the first trip echoes. As a result, there may be a discontinuity in the velocity field. This is another way to recognize the presence of multitrip echoes.

A second way that strange velocities can be detected by a radar is by having the radar beam reflect off of a nearby building or other object and detect a storm in some other direction. The returned echo will have the velocity associated with the storm detected, but it will be displayed in the direction the antenna was pointing when it hit the building. Sometimes this gives very clear discontinuities in the velocity field. Unfortunately, there is no easy solution to this except to move the radar to a different location (or remove the cause of the reflection). Fortunately, once the sources of these reflections are identified, they will always occur at the same azimuths, so recognizing them in the future becomes easier.

Color Fig. 1 shows some examples of this problem with the UND radar but from ground echoes instead of storm echoes. If you look carefully at the velocity data you can see some narrow spikes of zero velocity data at the edge of the nearby ground clutter which are pointed toward the east northeast and southeast; these are also detectable in the reflectivity data. If you go out to 80 to 100 km range along the same azimuths, some more weak reflectivity/zero velocity lines are shown; at this range there is also another such line at about 185° azimuth. Color Fig. 6 shows this same problem with slightly greater magnification.

All of these echoes appear to be caused by the radar signal being reflected off of buildings at nearby farms. This was confirmed by climbing the radar tower at South Roggen, Colorado, where the radar was located at this time, and looking in the directions at which these spikes were detected. In all cases there was a farm within a mile or two of the radar along these exact azimuths. Each of the farms had several buildings which could have caused the reflections. The nearby spikes are visible because the terrain rises gently toward the east and acts as a natural clutter barrier; little real ground clutter is seen toward the east. Consequently, the nearby buildings reflect the radar signal back toward the west where the local ground clutter extends farther away.

In the case of the echo at 80 to 100 km, the target detected by the radar was the Rocky Mountains! That is, the radar antenna would be aimed, for example, toward 70° azimuth; the transmitted pulse would hit a building a mile or two away, be reflected back towards the west, and detect the very strong ground targets 80 to 100 km away. The echoing signal would follow the reverse path and be displayed on the radar at the a range corresponding to the distance from the radar to the building plus the distance from the building to the mountains, but the azimuth would be that of the building. It is possible to detect storm echo in exactly the same way; in this case, however, the velocity would be the velocity of the part of the storm which is detected. In both cases, the reflected echoes make very interesting artifacts in the radar data. Beware of artifacts in your data!

A third way that incorrect velocity information can be displayed on the radar display is caused by the rotation of the antenna and the imperfect antenna beam patterns that all radars have. The feed horn of a radar antenna is usually located out in front of the antenna reflector and aimed back toward it. The electromagnetic radiation leaves the feed horn, hits the reflector, and is directed and focused out into space in the direction the antenna is pointing. There are two problems here. One is that not all of the energy hits the reflector. Some misses and goes more or less behind the radar while some of it goes off to the sides (actually, it goes in all directions, i.e., the beam pattern is really three dimensional). If the sidelobe energy hits a strong reflector, it can produce an echo at the range of the target, but it is displayed in the direction the antenna is pointing. The second problem is that, if the radar antenna is rotating when it is transmitting its energy, the frequency of the energy leaving the antenna will be a combination of the transmitted frequency *and* the frequency shift caused by the velocity of the feed horn.

Most feed horns are located at least a short distance from the center of rotation of the antenna. The speed of rotation of the feed horn is the product of the angular velocity of the antenna (e.g., radians/second) and the distance from the feed horn to the pivot point. For radars with big antennas, S-band radars, for example, this distance can be fairly large. The FL-2 S-band radar operated by MIT Lincoln Laboratory has the feed horn about 19 ft from the center of rotation.[9] If this radar scans at a speed of 30°/s, the feed horn will be moving at a speed of 3.1 m/s. Faster scan speeds give correspondingly faster velocities.

Now, when a radar transmits a signal in the direction the antenna is pointing, the angular velocity of the feed horn is perpendicular to this direction, so it does not add any *radial* movement; the transmitted frequency is unchanged. When the energy leaves a sidelobe, however, it will have a slight shift in velocity caused by the movement of the feedhorn. If this portion of the transmitted signal hits a stationary

---

[9] The Lincoln Laboratory FL2 radar (converted to C-band) was the prototype of the TDWR radar now in use by the FAA (see Appendix D).

ground target such as a building, it will appear to the radar that the building is moving at the velocity the feedhorn is moving. This produces an echo on the radar display that has the velocity of the antenna relative to the target.

Many very strong reflectors can produce sidelobe echoes which completely or nearly completely surround a radar. The correct position of this echo will be found when the antenna is pointed directly at it; here it will have its strongest reflectivity and a zero radial velocity. The apparent velocity of this target will vary sinusoidally as the antenna rotates around. It will have a maximum receding velocity when the antenna has rotated 90° away from it. At 180° rotation, the velocity will again be zero. As the feedhorn approaches from the other side, the velocity will then appear to be approaching the radar, with the maximum being reached at 90° from the correct position. If the antenna scans in the opposite direction, the approaching and receding sidelobe echoes will shift to the opposite sides.

Modern radar processing can reduce the effects of both range and velocity aliasing. Velocity folding can be reduced by applying algorithms to the data which look for velocity discontinuities with range and removing or reducing them by the Nyquist velocity. There are several schemes being developed for doing this. Automatic velocity unfolding is done with WSR-88D radar data so that user will not normally see any incorrect velocities.

There are also a number of schemes being tested to eliminate range aliasing. In some ways this is a more difficult problem, requiring more complex solutions. Some of the solutions involve changing the PRF in some clever ways to allow both long unambiguous ranges and high unambiguous velocities. These more complicated transmission schemes require more complicated processing to use them, but the results are worth the effort.

There are also automatic ways to select the PRF to give optimum coverage of a preselected location. For example, if we want to give warnings of microbursts over an airport, we want to select a PRF which minimizes the obscuration of the airport region by distant storms. Since these storms will move with time, the PRF will have to be changed to accommodate this movement. By examining the locations of echoes at long ranges (using low PRF's), it is possible to find the optimum PRF to give the least amount of obscuration in a nearby region of interest.

Multi-trip echo map projection

An alternative to changing the way the radar transmits or displays the radar data is to change the map on which the range-aliased data are displayed (Rinehart, 1989). By redrawing the map so that geographic and political points are distorted in the same way the second-trip echoes are, it is possible to determine the correct location for storms relative to the ground quite easily. Use of such a map projection makes it possible to tell what cities might be going to get rain soon or where a tornado might be right now.

An example of such a map display is shown in Fig. 6.6. The map on the upper left (Fig. 6.6a) shows the several-state region around the UND radar site near Denver, Colorado, during 1987 and 1988. The range circles are drawn at intervals of 150 km which is the maximum unambiguous range r_{max} corresponding to a PRF of 1000 Hz. Various weather stations are shown on the map. Figure 6.6b shows the area shown on a radar display set to cover 150 km range.

Figures 6.6c and d shows what the region would look like if all of the area within one or two r_{max} intervals, respectively, were eliminated and all locations farther away were shifted directly toward the radar by a distance 1 or 2 r_{max}. Obviously, the shape of the map becomes quite distorted.

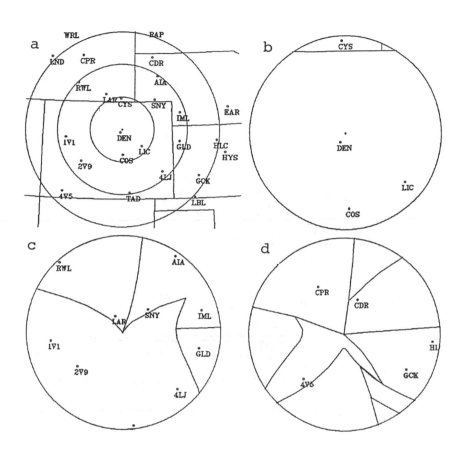

Figure 6.6   Maps showing the several-state region around the UND radar location at Denver, Colorado. Range circles on all panels are at r_{max} = 150 km intervals. By removing 150 km or 300 km from the center of the first display, second-(c) and third-trip (d) Rinehart projection maps are generated, respectively.

127

Chapter 7

Spectrum Width and Turbulence

When a radar detects a single target, the frequency shift in the returned signal is given by

$$f_d = \frac{2V}{\lambda} \qquad (7.1)$$

where f_d is the Doppler frequency shift, V is velocity, and λ is wavelength. When there are many targets within the sample volume (as there would be when a rainstorm is being measured, for example), each individual target would produce a frequency shift related to its radial velocity. The result is that a distribution of frequencies would be measured. Figure 7.1 illustrates a sample volume containing a small number of raindrops, each of which has its own speed and direction. In a real radar sample, there would be billions of raindrops present.

A Doppler radar usually processes all the returned signals to produce a single velocity for the entire sample volume. This is the mean velocity of the sample and is what we usually mean when we talk about the Doppler radial velocity.

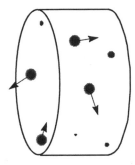

Figure 7.1 Illustration of a radar sample volume containing a few raindrops, each with its own speed and direction of movement. A real radar sample volume would contain billions of raindrops!

The way the mean velocity of a sample is determined depends upon how the Doppler processor was designed. Some compare signals from one pulse to the next pulse to determine velocity in what is sometimes called a "pulse-pair" processor. Another kind of processor transforms the time series data into the frequency domain using what is called a fast Fourier transform. Once the frequency data are available, it is a simple matter to determine which has the strongest component. In either case, the goal is the same: to determine the average velocity of the target.

One of the advantages of working in the frequency domain, however, is that it makes it possible to determine not only the average frequency but the distribution of frequencies as well. As mentioned, the frequencies in a single sample volume represent the velocities of all individual targets. By knowing what this distribution is, we can learn something about how variable the velocities are *within* a single sample volume.

This is useful meteorologically because it gives us information about turbulence within a storm.

There are a number of measures that can be applied to this internal, within-the-volume variability. The general term applied to it is called "spectrum width." It is, as its name implies, a measure of the width of the spectrum of frequencies detected (which, of course, are directly related to the distribution of velocities). Another term for this is the variance of the velocity, generally indicated using the symbol σ. Variance is essentially the average departure of individual velocities from the mean velocity of a sample and can be defined as

$$\sigma^2 = \frac{\sum (V_i - V_{ave})^2}{N-1} \tag{7.2}$$

where V_i is the velocity of an individual scatterer, V_{ave} is the mean velocity of all targets in the sample, and N is the number of targets in the sample. By taking the square root of the variance, we get σ which is really more of a standard deviation of velocity, having exactly the same units as velocity. σ indicates how variable the velocities are in the sample.

Sources of variance

The variance measured by a radar can be caused by a number of factors, not just the targets themselves. The variance contributed by each source is additive as follows

$$\sigma_v^2 = \sigma_s^2 + \sigma_a^2 + \sigma_d^2 + \sigma_t^2 \tag{7.3}$$

where the subscripts have the following meanings: v represents the total velocity variance from all sources; s indicates the variance caused by wind shear within the sample volume; a indicates variance caused by antenna motion; d represents the contribution due to the differential fall speeds of the different sized hydrometeors; and t represents the variance caused by turbulence. Let's consider these individually in at least a qualitative sense.

Variance from wind shear

Wind shear is a difference in wind speed and/or direction over a certain distance. For radar, this would be the difference in wind velocity between two points *within* a single sample volume. This can occur, of course, in three different directions: azimuthally, vertically, and radially. The total contribution to shear-induced velocity variance is the sum of the variances in each of the different directions.

Let's consider vertical wind shear for a moment. It is quite common in the atmosphere that the wind at one level will be different than that at another level (see Fig. 7.2). Winds in the jet stream, for instance, are often over 100 kt in speed; winds near the ground, even directly beneath the jet stream, are usually not more than 10 to 20 kt. Between the ground and the jet, the wind speed must change. This would be a case of vertical windshear. If a radar pulse volume samples different velocities at the bottom and the top of its volume, a velocity variance will exist due to this fact alone.

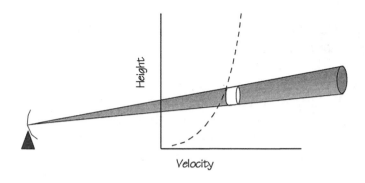

Figure 7.2   Windshear within a radar sample volume will cause targets in one part of the volume to have different velocities than those in another part.

Vertical wind shear strong enough to be detectable by a Doppler radar can occur fairly often at the interface between two different layers separated by a temperature inversion. At the inversion at the top of the planetary boundary layer there is frequently a shift in wind speed and direction. This occasionally gives rise to regions of moderately strong spectrum width. That this is the source of the signal may sometimes be verified by scanning in RHI mode instead of the more usual PPI mode.

Gust fronts are another source of shear-induced velocity variance. Gust fronts are distinct boundaries separating winds of two different velocities. They often show enhanced spectrum width because of the shear contribution alone.

Variance from a scanning antenna

The second major source of velocity variance in the equation above is that due to the scanning of the antenna. This is usually not a significant problem because most radars scan at a constant rate. Thus, even though scanning might contribute to the overall velocity variance, the contribution will be exactly the same everywhere within radar range. Since what we are usually looking for on a radar display is a region where the spectrum width is stronger than another place, variance produced by scanning will not mask real regions of stronger shear. It could, at least theoretically, make the meteorologically important shear appear somewhat stronger than it really is, however.

Variance from differential fall velocity

The third source of variance is that due to differential fall speeds of the hydrometeors. This effect is primarily important when we scan at moderately high elevation angles into a storm. This variance is given as

$$\sigma_d^2 = [\sigma_{d0} \, sin(\phi)]^2 \tag{7.4}$$

where ϕ is the elevation angle of the radar beam above the horizon, and σ_{d0} is the variability of terminal velocities from particles of different sizes in the same sample volume. The variance from σ_{d0} only contributes to σ_v when the antenna is aimed above the horizon. Since most meteorological radars do most of their scanning at nearly horizontal elevations, this term contributes little to most displays of spectrum width used by meteorologists.

Occasionally a Doppler radar will operate by scanning in a vertical direction. When this is done, the contribution to variance due to differential fall velocity is important and may, in fact, be the primary source of variance in the sample.

Measurements of the variance of hydrometeors were made during the 1950's and '60's with the following results. The variance due to rain is on the order of 1.0 m/s and is independent of the drop size distribution. For snow, σ_{d0} is on the order of 0.2 to 0.5 m/s, while for melting snow it is on the order of 0.7 m/s. For hail, σ_{d0} ranged from 8 to 25 m/s and is dependent upon both the hail diameter and the radar wavelength.

Variance from turbulence

The remaining term is perhaps the most important meteorologically. It is the contribution to velocity variance due to turbulence. It is the one which we will most often be interested in detecting with a radar.

In order for turbulence to be detectable by the radar, there must be a source of detectable echo. In some situations this is not a problem. Unfortunately, where we may want to see turbulence the most -- in clear air -- is where it may be undetectable. This is especially true high in the atmosphere. Turbulence near the ground, on the other hand, may often be detectable in what appears to be clear air because of the presence of insects, dust or other detectable debris picked up by the wind.

To summarize this discussion of variance so far, we can state that for low elevation angles and at constant antenna scan speeds, the variance of the radial Doppler velocity is predominantly a measure of the meteorological contributions, i.e., wind shear and turbulence.

It is possible to learn about turbulence on a somewhat larger scale from radars which cannot provide the full spectrum information. If we have the mean velocity at each point (as from a pulse-pair processor, for example), we can take the difference in velocity at consecutive points and determine wind shear from them. This gives a measure of shear or turbulence over distance on the order of a few hundred meters or so whereas the techniques discussed earlier gives information over distances smaller than this. For use by aircraft, distance scales of a few hundred meters is probably adequate.

There are a couple of airborne "turbulence" detecting radars on the market now which use this kind of approach in their processing. They examine the radial velocity at many points along the radar beam and determine where there is a change in velocity from one point to the next that is greater than a certain threshold. If such regions are detected, they are indicated on the radar displays as regions of turbulence. Such regions should be avoided because of the likelihood that they contain potentially dangerous levels of turbulence.

Other radar systems also process velocity differences in radial and azimuthal directions. NEXRAD, for example, has three displays for these products. One shows the shear or velocity difference from consecutive points in range; another shows the difference in radial velocities at adjacent points in azimuth; and a third combines these to show shear in both directions.

Chapter 8

Meteorological Targets

The ability to detect storms and other weather phenomena is perhaps one of the most valuable uses of radar. Severe storm and tornado warnings, hurricane observations, flood warnings and windshear warnings are all based on radar and result in the saving of lives and property each year. In this chapter we will explore how radar detects various weather events and some of the constraints and limitations in this activity. Since most of the meteorologically detectable echoes come from hydrometeors, we will begin with those and then consider other sources of echo that provide meteorological information.

Clouds

Clouds are usually not detectable by most radars but can, under some circumstances and with some radars, provide detectable echoes. For this discussion, let us distinguish between precipitating and non-precipitating clouds. If a cloud is not precipitating, either the particles within the cloud are too small to fall downward, the cloud is a new cloud that has not yet had sufficient time to produce precipitation-sized particles, or there is a constant, steady upward air motion that keeps the cloud particles suspended at approximately a constant altitude.

Clouds are composed of very small water droplets, ice crystals or both, depending upon the temperature and other factors. Many clouds that start out as liquid hydrometeors eventually change into either all ice clouds or a combination of both ice and supercooled liquid water droplets (i.e., liquid droplets whose temperatures are colder than 0°C; supercooled droplets can exist at temperatures between 0°C and - 40°C; liquid water does not exist below -40°C).

The sizes and concentrations of cloud droplets have been studied for many years. The size distribution within a cloud depends upon the kind, age and height within a cloud and the geographic location. In general, the farther from cloud base, the larger the droplets are. As a cloud gets older, droplets usually get larger. There is also a distinct difference between clouds that form over or near oceans and those that form well inland. Maritime clouds tend to have fewer droplets per unit volume than do continental clouds. Droplet sizes for both types range from perhaps 5 μm to 100 μm or more. Figure 8.1 (Fletcher, 1966) shows the mean droplet size distributions for various cloud types. Drop size distributions from other places and cloud types would differ from these in detail but would probably have similar general characteristics.

The radar reflectivity factor for clouds is generally quite weak. Since $z = \Sigma N_i D_i^6$, we can calculate reflectivities that might result from some clouds. For a continental cloud, the following table might be a reasonable size distribution:

Diameter (mm)	Number/cm^3	N D^6 (mm^6/m^3)
5	100	1.56·10^{-6}
10	100	1.00·10^{-4}
15	50	5.69·10^{-4}
20	25	1.60·10^{-3}
25	10	2.44·10^{-3}
30	5	9.19·10^{-3}
35	1	4.01·10^{-3}

$$\text{Total} = 1.80·10^{-2}$$
$$=> -17.4 \text{ dBZ}$$

The overall reflectivity of -17 dBZ is quite weak and would not be detected by most weather radars beyond a few kilometers; some radars could not detect echo this weak at any range. Weather radars on board most aircraft are not sensitive enough to detect clouds.

There is one interesting point from the above table which should be mentioned. This is the fact that the contribution to reflectivity from the small droplets, even though they outnumber the larger drops by one or more orders of magnitude, is generally negligible. Most of the reflectivity comes from the largest droplets. This is a consequence of the diameter-to-the-*sixth*-power term in the equation for reflectivity. The same feature is true of raindrop size distributions.

Rain

Rain is very easily detected by most radars. Rain can come in a wide variety of intensities, from light drizzle (the "Oregon mist" of the west coast) to the near-blinding downpours in severe thunderstorms. The measurement of rain by radar is one of the more important quantitative uses of radar. In the following sections we examine some of the properties of rain and how it affects its detectability by radar.

Raindrop size distributions

Raindrop size distributions have also been studied extensively over the past 30 to 40 years. There have been a number of techniques developed to sample these distributions. One of the most-used techniques was a raindrop camera which photographed a volume of rain with enough resolution that individual raindrops could be measured. From these or other size distributions, rainrate (e.g., mm/h), liquid water content (e.g., g/m^3)

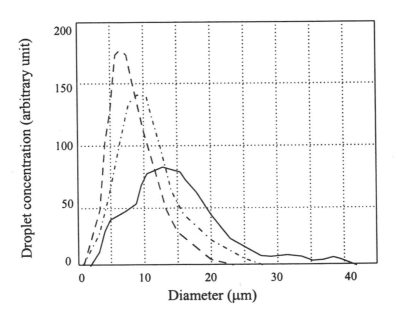

Figure 8.1 The mean droplet size distributions for various cloud types. Cumulus congestus (solid); altostratus (dash, dot, dash); stratus (dash). Based on Fletcher, 1966.

and radar reflectivity (mm^6/m^3) are easily calculated.

Figure 8.2 shows three raindrop size distributions collected at Ottawa and used by J. S. Marshall and W. McK. Palmer (1948); these are possibly the most widely known raindrop size distributions in all of meteorology, certainly in radar meteorology. They were the basis for the relationship known as the Marshall-Palmer distribution. This is a convenient relationship which gives an approximate size distribution for raindrops as a function of rainrate. As such, it is useful for various analytical exercises and for deriving other relationships.

The Marshall-Palmer relationship is given by the following:

$$N_d = N_O \, e^{-\lambda D} \qquad (8.1)$$

where $N_O = 8000/(m^3 \ mm)$, D is droplet diameter (mm), and λ is given by

$$\lambda = 4.1 \, R^{-0.21} \qquad (8.2)$$

and R is rainrate in mm/h.

Using this relationship and a specific rainrate, we can calculate the number of drops per unit volume and per unit drop size interval for any particular raindrop size. The size distribution can then be used to calculate the radar reflectivity or liquid water content of the rain.

Z-R relationships

From the above discussion it should be obvious that there is a relationship between rainrate and radar reflectivity. Experimentally measured drop-size distributions have been extensively used to calculate both radar reflectivity and rainrate. By plotting rainrate against reflectivity or by correlating these statistically,

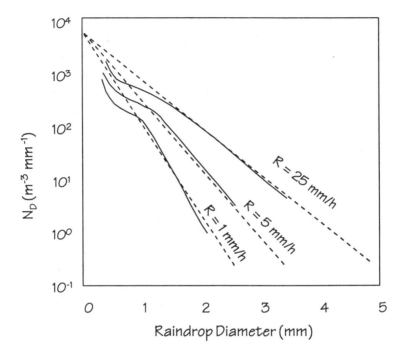

Figure 8.2 Marshall and Palmer drop-size distribution functions compared with the results of Laws and Parsons. Based on Marshall and Palmer, 1948.

we can determine the relationship between these two parameters. The most commonly used mathematical relationship between reflectivity and rainrate is the empirical relationship

$$z = A R^b \qquad (8.4)$$

where R is the rainfall rate (mm/h), z is the radar reflectivity factor (mm⁶/m³), and A and b are empirical constants. Battan (1973) lists more than 60 experimentally determined Z-R relationships; many more have been determined since 1973! While each applies to a specific time and place and rainstorm, there is often little difference from one to another. Unless a specific Z-R relationship exists for a given situation, there are only three or four that really need to be used. And even these are not dramatically different.

The most commonly used Z-R relationship is also due to Marshall and Palmer. It is $z = 200\ R^{1.6}$. This has formed the basis for much research and been widely used to calculate rainfall amounts from radar data. Indeed, as will be discussed in Chapter 10, radar is a very useful way to measure rainfall over large areas; Z-R relationships are the backbone of this activity.

DVIP levels

Radars can provide quantitative information on rainfall with excellent resolution. The radar reflectivity factor from rain varies from perhaps 20 dBZ (100 mm^6/m^3) to more than 50 dBZ (100000 mm^6/m^3). Reflectivities as high as 75 dBZ have been measured in storms, but reflectivities higher than about 55 dBZ are frequently associated with hail; the higher the reflectivity, the more likely it is that hail is present and the larger the hail is likely to be.

Radar signal processors can resolve moderately small differences in reflectivity. Many radars have a dynamic range (the difference between the strongest and weakest powers that can be detected, usually expressed in decibels) on the order of 80 to 90 dB; they frequently divide this range into 256 parts, giving a resolution on the order of 1/3 dB per measurement interval.

This much resolution is not always available, however. When digital signal processing was first applied to radar, it was not possible to produce nearly as many levels of intensity as is not taken for granted. Some of the early signal processor has only a handful of intensity levels.

The National Weather Service was able to divide the possible range of storm reflectivities into a relatively small number of intervals and get some very useful results. The processor used to do this was named the digital video integrator processor or DVIP. They divided storm intensities into six intervals but made the division on the basis of rainrates rather than radar reflectivity factor. Further, the divisions were made before the general change from English to metric units; rainrates in in/h were used for these divisions. Table 8.1 shows these intensity levels and the corresponding rainrates and reflectivities.

The same intensity levels used with NWS radars are used in a slightly modified form in aircraft radars. Radars on board aircraft use only four levels of intensity. These are levels 1, 2, 3 and 5.

Table 8.1   DVIP intensity levels based on rainrates and the corresponding radar reflectivity factors. Reflectivities are based on $z = 200 \, R^{1.6}$.

DVIP level	Rainrate (in/h)	Reflectivity (dBZ)
1	0.1	29.5
2	0.25	35.9
3	0.5	40.7
4	1.25	47.0
5	2.5	51.9
6	4.0	55.1

When examining a radar display using DVIP levels, be sure to recognize that if an echo shows at a given level, it means that the intensity shown is at least that level, probably stronger. For example, if an echo shows with a DVIP level of 3 on an aircraft radar, it means that the storm has a rainrate of at least 0.5 in/h ($Z \geq 40.7$ dBZ) but less than 2.5 in/h ($Z \leq 51.9$ dBZ). As a matter of safety, it would be prudent to assume that the detected reflectivity or rainrate is closer to the upper limit (i.e., the next DVIP level) rather than near the lower limit.

Snow

On American television weathercasts, it is quite common in the wintertime to hear forecasters say that radar doesn't detect snow very well. In some ways they are right, I suppose, but in another way they may be laying the blame on the radar instead of where it belongs, on the storms. Snow storms try to hide from radar as much as possible, it seems. Let's see why this is the case.

In reality, snow is easily detectable by radar. But there are some important differences between snow and rain, however. One of the major differences is that the precipitation rate for snow is usually much less than it is for rain. This comparison is based on the "water equivalent" precipitation rate. That is, the rate of snow is sometimes converted to liquid and then measured similarly to rainfall in mm/h of melted water. The amount of moisture that the atmosphere can hold is a very strong function of temperature; at warm temperatures the atmosphere can hold much more water vapor than at cold temperatures. One consequence of this is that the heaviest snowfalls (and rainfalls) occur at the warmest temperatures. The heaviest snows often fall when the temperature at the surface is just above the melting temperature of ice, i.e., 33° to 36°F. Of course, the temperature above the surface is colder than this; otherwise the precipitation would fall as rain instead of snow.

The second major difference between snow and rain is that the dielectric constant of ice is less than the dielectric constant of water. The $|K|^2$ term in the radar equation for beam-filling meteorological targets has a value of 0.93 for water but only 0.197 for ice. [Note that both of these numbers depend slightly upon radar frequency and/or temperature; for most meteorological radars and temperatures, we can ignore these variations.] Because of this difference, the power received back from snow and ice is about 7 dB less than it would be if a radar were looking at liquid precipitation. While ice crystals are usually larger than cloud droplets, ice crystals appear to a radar as though they are solid ice spheres of the same mass (Battan,

1973). Consequently, their larger size is not as great an advantage as it would seem because their density is usually much lower than that of pure water or solid ice.

Let's consider a numerical example. Color Figures 13 and 14 show two snow situations near Denver, Colorado. Radar reflectivities for these storms were on the order of 20 to 25 dBZ. For the WSR-88D radar, the radar constant from Appendix D is 64.9 dB; the minimum detectable signal power is -113 dBm. Equation 5.19 can be used to calculate the maximum range at which an echo of 20 dBZ can be detected. Equation 5.19 is

$$Z = C_3 + P_r + 20 \, log_{10}(r)$$

Substituting in values and solving for range r gives r_{max} = 2540 km. That is much farther than the tallest thunderstorm has ever been detected! So, low reflectivity is *not* a reason why snow cannot be detected! But echo height is.

The primary reason snow is not always detectable by radar is the shallow height of typical snow storms. Snow storms are usually much lower than most rain storms, especially the very tall thunderstorms that produce rain and hail. Snow storms are often very widespread in area but may only extend a few thousand meters above the surface. A storm that is 1500 m high would be below the radar beam at distances beyond about 120 km (assuming that standard refraction applies and that the radar used a minimum elevation angle of 0.5°).

All of these effects tend to make snow less easily detectable than rain. Whereas rain can usually be detected to very long ranges with a radar, snow is usually not seen out to the maximum range of a radar.

Chapter 8

Bright band

It is a fact of meteorology that much precipitation forms through an ice process rather than as an all water process. Much of the rain that falls to the ground begins as ice or snow. In the transition from snow to rain, some interesting changes take place that have consequences on what a radar sees.

As mentioned before, the reflectivity from ice is less than that from water for particles of the same diameter (or approximately the same mass). There is one other difference between snow and rain that must be mentioned, and this is terminal velocity. The terminal velocity of a freely falling object is the constant velocity that occurs when there is a balance between the force of gravity pulling it downward and the aerodynamic drag acting to slow it down. The terminal velocity of a particle depends upon its density and shape as well as the density and viscosity of the atmosphere. Spheres and other smooth objects fall faster than rough objects (of equal mass). Dense objects fall faster than light objects (of equal size). Objects fall faster high in the atmosphere where density is less than near the earth's surface where atmospheric density is higher.

So, with that as background, what happens when snow falls and melts, becoming rain? Above the melting level in the atmosphere (i.e., above the 0°C isotherm), snow will fall at a relatively slow terminal velocity. As soon as it reaches the melting level it will begin melting. Since the snow is falling through temperatures slightly above freezing temperatures, it will start to melt but from the outside toward the inside. This means that the extremities of the snow will melt first. When enough melting has taken place, the snowflake will have started to develop a water coating while still remaining moderately large and irregularly shaped. Thus, to the radar a melting snowflake will start to look like a large, slowly falling water droplet. Since the power received by the radar is proportional to $|K|^2$, the change from ice to water initially increases the reflectivity by as much as 7 dB.

As the water-coated snow continues to fall and melt, its size decreases and its terminal velocity increases. A consequence of the first effect is that the reflectivity decreases somewhat, depending upon the change in effective diameter between the snow and the water drop. The results of the second effect is that the drops leaving below the melting level move faster than those coming into it. This decreases the number density or concentration of snowflakes somewhat, and this further decreases the reflectivity in this region.

If we combine all of these effects, the reflectivity has the following characteristics. (See Figs. 8.3 and 8.4.) If we start with a given reflectivity in the snow above the melting level, there is on the order of 5 to 15 dB increase in reflectivity from the snow to the maximum signal received. Below this maximum the reflectivity will decrease 5 to 10 dB. In any case, the reflectivity below the maximum is usually higher than it was in the snow above.

The appearance of this phenomenon on a radar display depends upon the kind of display being used. On an RHI, it would be a layer of enhanced reflectivity just below the melting level. On older analog display tubes, this level was usually brighter than the other regions; hence the name "bright" band. On modern displays that show reflectivities using color, the bright band would be a colored band of higher reflectivity.

Chapter 8

On a PPI display, especially for a radar located at ground level and a bright band somewhere above the surface, it is necessary to tilt the antenna up somewhat so the beam will intersect the bright band at a shallow angle. Then, as the antenna scans around in azimuth, the radar will display a ring-like region of enhanced reflectivity. The melting level will be at the height corresponding to the most distant part of the bright-band region.

Bright bands occur primarily during stratiform or stable situations. When strong convention is present, the same physics apply, but the transition between snow and rain is often so chaotic as to be undetectable most of the time. When echoes are widespread and acting more sedately, it is much easier to detect bright bands. During the decaying portions of thunderstorms, however, bright bands will often be detected; their presence is usually an indication that the storm (or at least that portion of the storm containing the bright band) is dying out.

I was surprised to observe bright bands in decaying thunderstorms in Kenya while working on a hail-suppression cloud-seeding project. When a bright band became evident in a storm, it usually meant that the storm was no longer a hail threat to the region; operations for the day usually ended about the time the bright band appeared.

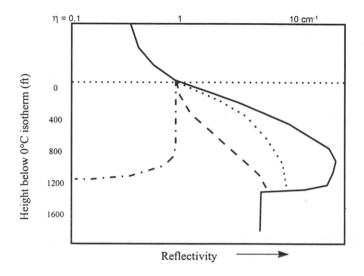

Figure 8.3 Schematic drawing showing the effects of particle coalescence (dotted), melting (dash), and changes in the terminal velocity (dot-dash-dot) on radar reflectivity through the bright band. From Austin and Bemis, 1950. Zero height is the freezing level. Radar reflectivity η is given along the top of the figure.

Figure 8.3 Schematic drawing showing the effects of particle coalescence (dotted), melting (dash), and changes in the terminal velocity (dot-dash-dot) on radar reflectivity through the bright band. From Austin and Bemis, 1950. Zero height is the freezing level. Radar reflectivity η is given along the top of the figure.

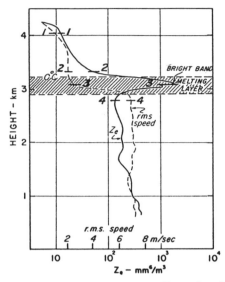

Figure 8.4 Simultaneous profiles of reflectivity factor Z and root-mean-square particle fall speeds in light (1 mm/h), steady precipitation with a bright band. Based on Lhermitte and Atlas, 1963.

Hail

Hail is defined as precipitation in the form of ice that has a diameter of at least 5 mm. It almost always occurs in thunderstorms but can fall from rainstorms that do not produce lightning and thunder; this is moderately rare, however. On the other hand, many thunderstorms produce lightning and thunder but no hail. Some people estimate that 85% of all thunderstorms contain hail at least during part of their lives.

Hail ranges from 5 mm to about 10 cm in diameter. On the other hand, the world's record hailstone that fell near Coffeeville, Kansas, on 14

September 1972, was 14 cm across it's longest dimension. Really large hailstones are rather rare, however, so they don't usually occur in large numbers.

Just as cloud droplets and water droplets have different sized particles present at one time, hail also falls with a size distribution that depends upon the storm that produced it. Because hail can vary so much from the smallest to the largest stones and stones fall at velocities that depend upon their sizes, it is not unusual for the largest stones to fall out first, followed by smaller and smaller stones. The gravitational sorting that produces this along with the fact that most hailstorms are moving along at moderate speeds can combine to make one point on the ground have large hail while nearby locations will have much smaller hail.

The terminal velocity of hail, as mentioned, depends upon the hailstone diameter. It also depends upon the shape of the hail (i.e., its "drag coefficient") and on the density of the air. Measurements and/or calculations of hail terminal velocity have usually found that hailstone terminal velocity can be expressed in a power relationship of the form $V_t = A\ D^{0.5}$, where D is hailstone diameter (usually in cm) and V_t is in m/s; A is an empirical constant. One set of measurements found a value of 11.45 for A (Matson and Huggins, 1979). This applies at ground level; higher in the atmosphere where air density is lower, the terminal velocity would be proportionally higher.

The reflectivity from hail depends upon whether the outside surface is wet or dry or if there is any water enclosed in the hail (i.e., spongy hail). Dry hail has a lower reflectivity than wet hail of the same size. This is again a consequence of the different dielectric constants of ice and water. The reflectivity from a single hailstone (or collection of stones falling through similar conditions at the same location) can change as it falls from above the melting level to below the melting level.

A final complication for hail is that it is often large enough that Rayleigh scattering conditions do not apply. That is, the hailstones are in the Mie region. For 3-cm and 5-cm wavelength radars, almost all hail is in the Mie region; small hail detected by 10-cm wavelength radars would still be in the Rayleigh region, but large hail would be in the Mie region. Chapter 10 discusses the use of dual-wavelength radar (i.e., a set of two co-located radars with different wavelengths) specifically for the detection of hail.

One consequence of having hail in the Mie region is that the backscattering cross-sectional area of a hailstone can actually increase as it melts and gets smaller. This effect might not be detectable with a radar, however, because the radar pulse volume is usually so large that hundreds if not thousands of individual hailstones will be contributing to the received power at the same time; the effects of an individual stone would be less important because of the large number of stones present.

Attenuation

Electromagnetic radiation passing through any medium is reduced in power by an amount that depends upon the kind of material present and its density. Some materials reduce or attenuate the radiation more than others. In free space where there is no material (as in the nearly empty space between the earth and the moon, for example) there is no attenuation; anywhere within the atmosphere there has to be at least a little attenuation. Because attenuation can have such important consequences on the use of radar, let's examine it and its causes in more detail.

Atmospheric Attenuation

The cloud- and precipitation-free atmosphere still contains nitrogen, oxygen, water vapor, and other gases in lower amounts. Nitrogen and many other gases cause no significant attenuation at radar wavelengths. Oxygen and water vapor do. Figure 8.5 (Bean and Dutton, 1968) show the attenuation of both oxygen and water vapor as a function of frequency. From this figure it is obvious that attenuation is not much of a problem at frequencies below about 10 GHz. However, when the water vapor is higher than the amount used for the figure, the attenuation will be higher.

Note that the attenuation given along the ordinate is in dB/km. An attenuation of 0.01 dB/km over a 100 km path will produce a total of 1 dB of attenuation. Also, since radar works by transmitting a signal and receiving an echo back, the path traveled by the radar waves will be twice this distance, producing for our example a total of 2 dB of attenuation.

Because of earth's curvature, the height of a radar beam will usually get higher as the distance from the radar increases. Most of the attenuation suffered by radar waves will thus be close to the radar. Figure 8.6 shows the attenuation for two-way radar propagation as a function of radar frequency for elevation angles of 0° and 5°.

In conclusion, the attenuation caused by the atmosphere is usually quite small and is often neglected. If extremely accurate measurements are needed, atmospheric attenuation can be corrected for by increasing reflectivities as a function of range and elevation angle.

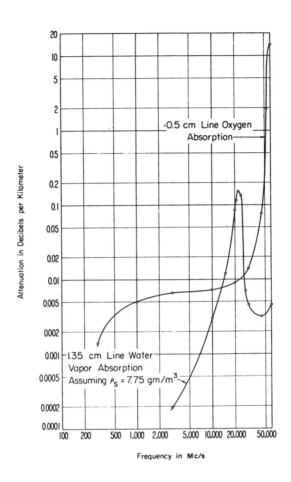

Figure 8.5 Atmospheric attenuation from water vapor (dashed curve) and oxygen (solid curve) at standard pressure (1013.25 mb) as a function of frequency. The water vapor curve assumes an absolute humidity of 7.5 g/m³. From Bean and Dutton (1968)

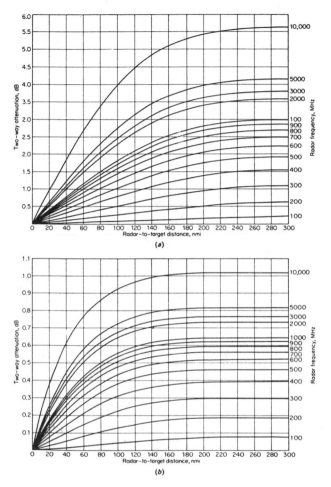

Figure 8.6   Attenuation for two-way radar propagation as a function of range and frequency for elevation angles of 0° (a) and 5° (b). From Skolnik, 1980, Introduction to Radar Systems, with permission of McGraw-Hill, Inc.

Cloud Attenuation

Attenuation by clouds is considerably more variable than that from the atmosphere because clouds themselves are more variable, ranging from nonexistent to very thick clouds. Further, it depends upon whether the clouds are composed of water droplets or ice particles. Except for very long paths through ice clouds, the attenuation through ice clouds is probably negligible.

Table 8.2 gives the attenuation rates for clouds as a function of radar wavelength, cloud temperature, and whether it is water or ice. For the ice situations the attenuation rates range from about 0.0006 to 0.009 $(dB/km)/(g/m^3)$. Again, we can see that there is more attenuation at higher frequencies (shorter wavelengths) than at lower frequencies. Also, the low attenuation rates for ice clouds are clearly evident.

Table 8.2. One-way attenuation coefficient K_1 in clouds in $(dB/km)/(g/m^3)$. From Gunn and East, 1954.

	Temp. (°C)	Wavelength (cm)			
		0.9	1.24	1.8	3.2
Water	20	0.647	0.311	0.128	0.0483
	10	0.681	0.406	0.179	0.0630
	0	0.99	0.532	0.267	0.0858
	-8	1.25	0.684	0.34*	0.122*
Ice	0	0.00874	0.00635	0.00436	0.00246
	-10	0.00291	0.00211	0.00146	0.00081
	-20	0.00200	0.00145	0.00100	0.00056

* extrapolated

For water clouds, the amount of attenuation cannot be ignored for most radar wavelengths if the clouds are at all dense and/or extensive. For example, a cloud with a liquid water content of 4 g/m^3, a temperature of 20°C and a one-way path length of 25 km using a 3.2-cm wavelength radar would have a total attenuation of 10 dB.

Rain Attenuation

As should be expected, attenuation by rain is even stronger than it is from clouds. Tables 8.3 and 8.4 give the attenuation rates for rain as a function of rainrate and radar reflectivity, respectively. For X-band radars, the attenuation rates are high enough that severe attenuation can occur in many rain situations, especially thunderstorms. For example, using the Mueller-Jones relationship at 3.21-cm wavelength, a 10-km one-way path through a thunderstorm having a 100-mm/h precipitation rate, the attenuation would be 11.6 dB. Longer paths and/or heavier rainfalls would produce even more severe attenuation.

As another example, if an X-band radar detects a storm having a DVIP level of 6 (using Z = 55.1 dBZ) of 10-km extent, it would also produce an attenuation of 11.6 dB. This is probably not enough attenuation that a weak or moderate storm beyond the first storm would go undetected, but it is enough that the distant storm's intensity would be significantly underestimated.

Snow Attenuation

While snow causes more attenuation than clouds, the total amount of attenuation caused by snow is usually negligible. The low attenuation in snow is a result of the same factors discussed for general snow detection, namely, the dielectric constant effect, the lower melted-precipitation rates in snow as compared to rain, and the generally low clouds that produce snow.

Table 8.3   One-way rain attenuation K' in
(dB/km)/(mm/h). From Wexler and Atlas, 1963.

λ. (cm)	M-P (at 0°C)	Modified M-P (0°C)	Mueller-Jones (0°C)	Gunn and East (18°C)
0.62	0.50-0.37	0.52	0.66	
0.86	0.27	0.31	0.39	
1.24	$0.117R^{0.10}$	$0.31R^{0.07}$	0.18	$0.12R^{0.06}$
1.8				$0.045R^{0.11}$
1.87	$0.0045R^{0.10}$	$0.050R^{0.10}$	0.065	
3.21	0.005-0.007*	0.0053	0.0058	
4.67	$0.011R^{0.15}$	$0.013R^{0.15}$	0.018	$0.0074R^{0.31}$
5.5	0.003-0.004*	0.0031	0.0033	
5.7				$0.0022R^{0.17}$
10	0.0009-0.0007*	0.00082	0.00092	0.0003

*First value applies at 2 mm/h, second at 50 mm/h, and there
is a "smooth transition" between them.

Table 8.5 gives the attenuation rates for four
different wavelengths and three different precipitation
rates. A 50-km two-way path through snow at 10 mm/h
precipitation rate would produce only 2 dB of
attenuation. For all but precise measurements, this
could probably be ignored.

Table 8.4   Attenuation by rain expressed in terms of z (mm^6/m^3). Except for those from McCormick, the values are based on the modified MP data in Table 8.3 and a Z-R relationship of z = 300 $R^{1.5}$.

Frequency (GHz)	Wavelength (cm)	k_p (dB/km)
15.0	2.0	$7.15 \cdot 10^{-4} z^{0.725}$*
9.3	3.21	$1.18 \cdot 10^{-4} z^{0.67}$
8.0	3.75	$1.16 \cdot 10^{-4} z^{0.806}$*
5.5	5.5	$6.9 \cdot 10^{-5} z^{0.67}$
3.0	10.0	$1.83 \cdot 10^{-5} z^{0.67}$

From McCormick, 1970

Table 8.5   One-way attenuation coefficients (dB/km) by low-density snow at $0°C$ calculated from: $k_s = 3.5 \cdot 10^{-2} R^2/\lambda^4 + 2.2 \cdot 10^{-3} R/\lambda$ (Battan, 1973).

Wavelength (cm)	Precipitation rate R (mm/h)		
	1	10	100
1.8	0.0046	0.344	33.5
3.2	0.0010	0.040	3.41
5.4	0.00045	0.0082	0.45
10.0	0.00022	0.0026	0.057

Hail Attenuation

It is much more difficult to quantitatively estimate the attenuation from hail. Hail is quite variable in duration, extent and intensity. In reality, hail is such a rare atmospheric phenomena that it is usually not present at all. When hail is present in a storm, it usually accompanies rain, and often very heavy rain. The net effect of both rain and hail is to produce even more attenuation than without the hail.

Correcting for attenuation

In the preceding sections we found that it is possible to estimate the amount of attenuation that might be taking place if we know the conditions (i.e., rainrate or reflectivity) correctly. Armed with that knowledge, can we then correct for the lost signal? The answer is an unequivocal "maybe" or "sometimes"!

With the exception of atmospheric attenuation by gases, we usually don't really know the total extent of the area that is causing attenuation. Gas attenuation is quite straightforward. Given the path of the radar signal, we should be easily able to calculate and correct for gas attenuation. Even water vapor, which is quite variable in time and space, can often be determined sufficiently accurately that we can correct for losses through it. And further, the losses from these are typically fairly small and are often ignored.

Correcting for attenuation in cloud and precipitation is not as easy. Clouds are typically not detectable on radar, yet the attenuation in them may exceed that from gases in the atmosphere. Precipitation, on the other hand, is usually clearly displayed on a radar, at least to the far edge of the storm *as seen by the radar!* And therein lies the problem. Once the attenuation exceeds a certain amount such that no signal is coming back through a storm, more attenuation ·will only reduce the maximum range of echo detection in that direction. So, what the radar sees on the far side of the storm may have already been affected by attenuation. As long as there is some echo detected, it may be possible to estimate how much stronger the echo would have been if there had been no attenuation between the radar and that point. Attempting to correct for attenuation can give very incorrect estimates of reflectivity or rain rate. Be careful if you try this that your final estimate doesn't start to approach infinity!

For certain purposes, however, it might still be important to try to correct for attenuation, provided enough is known about the situation to do so safely. Rainfall estimates, for example, could be improved if the attenuation were accurately known and accounted for. There are techniques to estimate the echo lost from attenuation. But once the echo is completely gone, it is impossible to recover anything beyond that point unless some other assumptions or information is available to fill in the missing information. For example, having another radar (or two or three) looking at a storm from another direction can provide sufficient information to correct for attenuation losses. This is essentially what is done when data from a network of radars is combined by using the strongest return at any point from any of the radars in the network.

Recognizing the presence of attenuation

Can you recognize when attenuation is a problem with a radar? Sometimes. It depends upon the kind, intensity and extent of the precipitation being detected. During thunderstorms or very strong echoes, it is almost certain that some attenuation will take place.

Sometimes it is obvious that attenuation is a problem. When very strong attenuation takes place there will often be a radial region on the far side of an echo that contains very weak echo or none at all. This attenuation "shadow," as it has been called, is evidence that the storm producing the attenuation is very that the storm producing the attenuation is very intense. It may not be clear, however, that the region on the far side of the storm is attenuated. Some innocuous storms have shapes that might naturally produce the same kind of pattern. How can you tell the difference? From a ground-based radar it may be difficult if there are no storms on the far side of the nearest storm. If more distant storms are present, they may give a clue about what the nearer storm is really doing. Then again, maybe they won't.

Color Figure 11 shows a region of attenuation on the far side of the strong, tornado-producing thunderstorm 20 to 40 km to the southwest of the UND radar. The reduced signal is evident in both the reflectivity and the velocity portions of this figure. It is difficult to estimate the exact magnitude of the signal loss, but it is not difficult to see that some of the echo is missing. By comparing the reflectivity of the clear-air echo in the general area with that being detected beyond the storm, it would appear that the tornadic storm produced as much as 10 to 15 dB of attenuation, and possibly more.

In an airplane it is possible to operate a radar to improve your chances of recognizing attenuating situations. To do this, you can tilt your antenna down so that the ground is being detected at a range near the maximum range of the radar. Then when a storm exists between the aircraft and the normal maximum range of the radar, the absence of any ground clutter behind the nearby storm would be clear evidence that attenuation is present. This technique has been suggested by Archie Trammell (1989) in his training courses on the use of airborne radar. His suggestion of flying with the antenna pointed just low enough to give ground clutter at a range near the maximum range of the radar is an excellent way to insure that your radar is working properly as well as giving warning of strongly-attenuating situations.

Other Meteorological Targets

Besides those already discussed, there are certainly other targets of meteorological interest. Among those not yet mentioned are tornadoes, hurricanes, mesoscale convective complexes, and various wind phenomena. Wind phenomena will be discussed in the next chapter on clear air targets.

Ever since Don Staggs of the Illinois State Water Survey detected the first tornado with a radar in 1953, tornadoes have been detected and extensively studied by radar. While tornadoes are quite rare at any given point on earth, in the central United States it is not at all uncommon to have tornadoes *within radar range* of a single radar almost every year. Tornadoes are so prevalent within the United States, in fact, that the WSR-88D (NEXRAD) radars were given the capability to measure Doppler velocities primarily so they would be better able to detect tornadoes. NEXRAD algorithms are being used specifically for the detection of mesocyclones and the tornadoes themselves.

Color Figure 11 shows an example of a tornado vortex signature (TVS) detected by the UND C-band radar in the Kansas City, Missouri, area. The TVS is the small region located at approximately 35 km range and 215° azimuth. At that point there is a very small region of quite variable velocities associated with a tornado. On the reflectivity image there is an appendage of echo off on the southwest side of the main storm which, while not being the perfect example of the classic "hook echo", does have hook-echo characteristics. As mentioned earlier, however, this is also on the edge of a region where attenuation is taking place, so that may contribute to the apparent shape of this echo.

Hurricanes are also amenable to study by radar -- provided they are close enough to the radar to be detectable. Hurricanes generally form hundreds or thousands of kilometers away from land. Ground-based radars have detectable ranges on the order of 400 km or so. Consequently, hurricanes are not detected until they have moved close enough to land to already be a problem. On the other hand, airborne Doppler weather radars have been used very effectively in studying hurricanes (a case of Muhammad going to the mountain instead of waiting for the mountain to come to him!). Once a hurricane gets within radar range, radar can be very useful (if it survives the storm!) in measuring the copious amounts of rain that often fall during hurricanes. Also, many hurricanes spawn one or more tornadoes shortly after they make landfall; these can also be detected and warnings issued using the aforementioned detection algorithms.

Figure 8.6 is one of my all-time favorite hurricane pictures from a radar. In fact, it is the very last full-circle image collected by the Miami WSR-57 radar before the radar was destroyed by Hurricane Andrew in 1992! The picture clearly shows the eye of Andrew as it move onshore near Homestead Air Force Base in Florida. This radar has since been replaced by a new WSR-88D radar.

Mesoscale convective complexes are, as the last word in the name implies, complexes of numerous storms acting as a giant unit. As such, the various components are just as detectable as they would be if they occurred individually. MCC's are sometimes so large, however, that a single radar does not have the ability to cover the entire event; they are just too extensive. Data from more than one radar are routinely combined, fortunately, so that entire events can be monitored by ground-based radar networks. The NEXRAD network will add velocity measurements to the reflectivity measurements already available and should provide even more useful information to study these events.

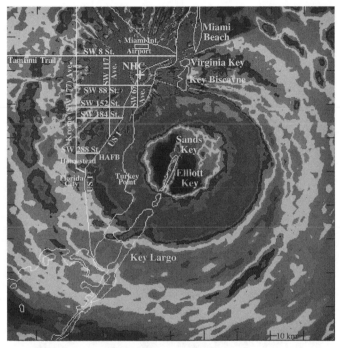

Figure 8.6 PPI from the Miami, Florida, National Weather Service WSR-57 radar just before Hurricane Andrew destroyed the radar.

R. E. Rinehart, Marshal Hagen and Terry Krauss, 1997

Chapter 9

Clear-Air Return

Meteorological information can come from nonmeteorological as well as meteorological targets. When we think of weather radar, we usually think in terms of the radar detecting echoes from weather. However, since radars can receive detectable power from insects and other targets, it is often possible to learn about the weather from these nonmeteorological targets. In fact, as will be seen, some important wind phenomena are detectable largely because of clear-air echo.

As a historical note, the detection of echoes in the optically clear air began fairly early in the use of weather radar. Since there was nothing visible to the human eye or even through binoculars and telescopes, these unidentified echoes were given the names of "angels" or "ghosts." Angels were discrete, point targets (most likely birds, in many cases) while those that were more nebulous and diffuse and seemed to cover an area were called ghosts. There are many papers in the early history of radar meteorology related to angel and ghost echoes. There were a number of experiments in the early 1960's in which individual birds and insects were tracked by radars, demonstrating that these were likely the source of most of the unknown echoes. The last paper in the radar conference preprint volumes using "angel" in the title was published in 1970. There was even one paper published on an unknown echo detected over water; it was dubbed a "mermaid"!

Sources

There are two general categories of radar echo in the clear air. The first is from insects, dust, chaff and other particulates in the atmosphere that are large enough to return some power to the radar. The second source of return is from what is called refractive-index gradients. Let us consider each of these individually.

Particulates

The return from particulates in the atmosphere is usually very similar to the return from raindrops or cloud particles. The kind of particulate present depends upon the weather itself. When it is windy enough to cause blowing dust and debris, the concentration and size of these particles can be enough to give a radar-detectable echo. Normally, however, the wind is not nearly this strong, so this would occur only for storm situations where strong winds occur.

Examples of radar-detectable dust clouds occasionally occur. Silas Michaelides related one such incident that took place in December 1996. A large echo moved across the Mediterranean Sea, across Cyprus, and toward the northeast to Syria and Israel (personal communication, 1996)

Much of the time the winds in the atmosphere are much lighter than necessary to cause blowing dust. Under these circumstances, there may still be detectable clear-air return if there is a sufficient number of insects present. This occurs over much of the world during the warmer months of the year.

Insects are found in the atmosphere from the surface to moderately high altitudes. Of course, since the source of all insects is at the surface of the earth, the number of insects decreases quite rapidly with height. However, insects may be in concentrated layers during periods of migration. The height of these layers is dependent upon the type of insect, the temperature and the wind. Radar is so useful for the study of insects that there is a related scientific field of research called radar entomology (Clark, 1997).

If the insects detected by a radar are moving in random directions or if the speed of the insects is slow compared to that of the wind, the wind speed and direction in the atmosphere can be determined from the return from insects. Even during migrations, most insects probably fly at only a couple of meters per second relative to the air. Thus, when wind speeds are much stronger than this, the dominate signal is due to the wind and not to the speed of the insects. But be careful. The measured velocity when insects are present is always a combination of the wind and the insects' velocities.

One of the best uses of the radar echo from insects is to estimate the winds in the lower troposphere. By scanning a Doppler radar a full 360° of azimuth with the antenna pointed above the horizontal (or repeating 360° scans at several elevation angles), the radar data can be used to provide wind data at various heights in the atmosphere. A technique called VAD (velocity-azimuth display) analysis can be performed which determines wind speed and direction at any range from the radar (Lhermitte and Atlas, 1961). By knowing the elevation angle and range, we can determine the height of the radar beam. By performing this at different ranges and/or elevation angles, winds can be determined at various heights. Then, by making such measurements at specific time intervals, we can observe changes in the winds with time which indicate current and past conditions, making it possible to estimate future wind conditions. The basic VAD technique usually assumes uniform winds at a given level so that the signal is a pure sine wave. It is possible, however, to gain even more information by accounting for departures from pure sine wave shapes. See Fig. 11.2 For an example of a VAD wind profile collected by a WSR-88D radar.

As an example of clear-air return, consider the data shown in Color Fig. 2. These data were collected at 1947 UTC (1447 CDT) on 6 June 1989 with the UND radar at Kansas City, Missouri. On this particular display, there are no echoes being detected from clouds or storms. But there is some very useful weather information on the velocity display. The approaching velocities from the south and receding velocities toward the north clearly show that the wind on this day is from the south; the colors indicate that the wind is very light at the radar but reaches speeds as fast as about 10 m/s at the longer ranges (corresponding to the higher heights). The zero isodop is quite straight (oriented west to east), indicating that the winds being detected are all from the same direction. Since the data were collected at 0.3° elevation angle, however, even at the maximum range of 56 km, the beam is only 480 m above the surface, so we are really looking only at the very lowest portion of the atmosphere.

Notice on this pair of figures that there is also some strong echo present, especially from 15 to 30 km to the south. This is ground return from the buildings and hills near downtown Kansas City. Where the ground return is strong, the velocity display shows zero velocity. It is only where there is no detectable (or very weak) ground clutter that the radar is able to detect the clear-air return. Most of the clear-air return has reflectivities of about 6 dBZ or less.

Color Figure 11 is another good example of the detectability of clear-air return. Except for the very strong, tornado-producing echo to the south of the radar, almost all of the other echo is caused by clear-air return. In this case the reflectivity is as strong as 14 dBZ toward the northeast but is generally less than this elsewhere. Winds in the lowest layer are from an azimuth of about 215° at a speed which increases with altitude. Again, close to the surface (and radar) the velocity is very light. Toward the northeast near 56 km, however, speeds are aliased and going away from the radar at a speed of about 25 m/s.

This figure also has a second interesting wind feature. The zero isodop rather abruptly changes direction toward the northwest at about 20 km. Since the wind is always perpendicular to the zero isodop (unless, of course, it really is zero velocity, an uncommon feature in the real atmosphere, especially when there is clear evidence that the atmosphere is not at rest everywhere), we can see that the winds to the northwest are moving toward the radar from the northwest at a speed of about 15 to 18 m/s. This abrupt change in wind direction is evidence of an approaching wind shift of some kind or another, probably a cold front. Further, behind this wind-shift line there is noticeably less clear-air return. If the air behind the wind-shift line is cooler than that in front, it would be quite reasonable for the insect activity to decrease there.

If you look carefully at the other color pictures collected during warm months, you can detect regions of clear-air return on almost all of them. It would be the regions where there is weak reflectivity (here "weak" means something on the order of 10 to 20 dBZ or less) but where there is widespread velocity data showing. The source of this clear-air return is most likely from insects being carried along by the wind.

Clear-air echo is not always present. On Color Fig. 1, for example, there are some scattered pixels of reflectivity of about -2 dBZ but these points show no velocity data. In this case there is no clear-air echo being detected. The echo that is showing is simply receiver noise. By changing some of the thresholds in the system it is easily possible to eliminate this kind of echo. However, at least to some older-generation radar meteorologist, it is reassuring to see a little of the receiver noise to let us know that things really are working correctly.

An ambiguous source of echo

As another example of the return from clear air, consider Fig. 9.1 from 2233 UTC (1733 CDT) on 14 August 1989 at Kansas City. This scan was collected at an elevation angle of 6.7°. There is a very clear ring of weak echo near 25 to 30 km range with another weaker ring discernible near 18 to 20 km. Both of these rings are definitely clear-air echoes, but their exact source is not absolutely clear. As mentioned, there are two candidates for the source of these echoes, insects and gradients in the refractive index. It is not absolutely certain which of these is the source of these echoes.

This day was characterized by high pressure and subsidence in the Kansas City region. Kansas City's temperature and dew-point temperature were 82°F and 62°F (relative humidity of about 30%) at 00 UTC. A few cumulus clouds were reported (but none seem to be present in radar data shown in Fig. 9.1). Surface winds were 5 kt from the SW, increasing to about 10 kt from the SW at 850 hPa, and from the NW at 10 kt and 20 kt at 700 hPa and 500 hPa.

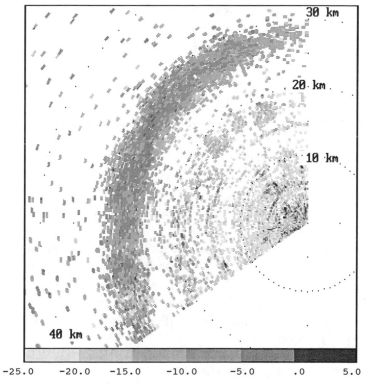

```
Radar: U. N. D.          Site:  Kansas
Date:  08/14/1989        Time:  22:33:04 UT
Field: DZ                Elevation:  6.7
```

Figure 9.1 Radar reflectivity factor data (dBZ) at 2233 UTC on 14 August 1989 at Kansas City (elevation angle of 6.7°).

The reflectivity in the highest layer of clear-air echo is on the order of -10 to -5 dBZ . The lower layer has reflectivities near -20 to -15 dBZ. The winds detectable by radar are in good agreement with NWS wind data.

So, what is causing the echoes that are being detected? This particular discussion is sandwiched between general discussions of clear-air return from particulates and from refractive index gradients. The honest answer to what is causing this return is that I am not sure! My inclination and prejudice makes me want to claim that the return is from insects. Indeed, I have seen other examples of elevated layers of clear-air return for which a reasonably convincing case can be made that they are caused by insects. This case is not so clear. Some people have claimed that when clear-air return is in an elevated layer, it is most likely caused by refractive index gradients.

Because the winds are generally light and without a lot of speed or direction shear, it seems unlikely that there would be a lot of turbulence present to cause gradients in the refractive index. The air is both dry and stable. On the other hand, the height of the upper layer -- about 12 kft above the radar -- seems a little high for insects. But it is summer, and insects are very common in summer. For this particular case it seems we cannot make any absolute claim either way without some additional information. This is sometimes the situation; we have to reserve judgment. [But personally, I still *believe* it is insects!]

Refractive index gradients

The second source of clear-air return in the atmosphere is from refractive index gradients. When the refractive index of the atmosphere changes significantly over distances which are small compared to the wavelength of the radar, the region containing these sharp, small-scale fluctuations can return some of the incident power back toward the radar. If the radar is sensitive enough, this can cause detectable echo to be displayed on the radar.

From our earlier discussion, we saw that the refractive index of the atmosphere depends upon the pressure, temperature, humidity, and free-electron concentration of the atmosphere (Eq. 3.5); the concentration of free electrons is important only in the ionosphere, not near the surface. Rapid changes of humidity or temperature can cause rapid changes in the refractive index. It is highly unusual, however, for the pressure to change significantly over very short distances, so pressure fluctuations are not a source for this kind of clear-air return. Temperature and humidity fluctuations can be strong enough in the atmosphere, however, to cause detectable echo on some occasions.

The primary cause of these small-scale fluctuations is turbulence. When turbulence the size of the radar wavelength (or smaller) mixes warm and cold air or dry and moist air (or of the right combination of both), the radar may be able to detect it. This scatter from turbulence in the atmosphere is sometimes called Bragg scatter. It is the source of much of the echo detected by the new generation of wind-profiling radars.

One of the important provisos in the introduction of this section, however, was the constraint that the fluctuations be small compared to the wavelength. This suggests that longer wavelength radars will stand a better chance of detecting refractive index gradients than short wavelength radars because there is a greater distance over which the atmosphere can produce temperature/humidity gradients. Indeed, the newer wind measuring radars (wind profilers) were designed to operate at wavelengths on the order of 0.3 to 6 m (rather than the more common 3- to 10-cm wavelengths used for storm and weather detection) so they would be able to detect winds by getting their signal from temperature and humidity fluctuations in the atmosphere.

Clear-air return from refractive index gradients is a fairly uncommon occurrence with most microwave radars, especially radar used on board aircraft. We will not discuss this further except to say that clear-air return from temperature/humidity gradients does exist and should be recognized as a source of some radar echoes.

Unfortunately, in the one case where it would be nice to detect clear-air return from an aircraft, i.e., clear air turbulence, there is not enough detectable signal with most airborne radars to make its detection operationally useful.

For a more complete discussion of Bragg scattering and the reflectivity caused by gradients of refractive index, see Gossard and Strauch (1983) and Doviak and Zrnic' (1993).

Detectable wind phenomena

The presence of detectable echo in the optically-clear atmosphere makes it possible for radar to detect certain meteorological phenomena. While the preceding section presented two sources of clear-air echo, it should be emphasized that the primary source of most clear-air return with conventional, microwave radars is usually from particulates, i.e., insects, rather than refractive index gradients. Because of this, the phenomena detectable by radar in clear air is usually limited to situations where insects are present, namely during spring, summer and fall; clear-air return during winter months is far less common than during the rest of the year (see Color Fig. 1).

Winds

The detection of wind has already been mentioned. The general wind field over a region is now routinely seen on many modern Doppler weather radars, at least in and around the radar site and to altitudes up to as much as a few hundred meters. But other wind phenomena are also clearly detectable with sensitive radars. Let us consider these now.

Microbursts

One of the most important wind phenomena to be detected by Doppler radars in recent years is the microburst. Since the recognition of microbursts as distinct meteorological events by Dr. Ted Fujita of the University of Chicago in 1974, microbursts have been studied in great detail. Since 1970, more than 500 people have been killed in aircraft accidents caused by microbursts (Fujita and McCarthy, 1990).

What are microbursts and what causes them? Microbursts are small but powerful downdrafts from clouds or rain storms that hit the ground and spread in all directions. Fujita defines a microburst as: "A small downburst with its outburst, damaging winds extending only 4 km or less. In spite of its small horizontal scale, an intense microburst could induce damaging winds as high as 75 m/s (168 mph)".

All convective clouds are caused by moist air rising, condensing, and forming cloud droplets. If conditions are right, precipitation will form, causing the cloud to turn into a rainshower or thundershower. The very existence of a convective cloud is visible evidence that there are rising air currents in the atmosphere. Since the atmosphere is a continuum, it is impossible to lift air at one location without forcing air to sink somewhere else. Sometimes this compensating downward movement is spread out over a large area around the cloud so that the downward velocity is quite small. In many clouds, especially the larger ones, the downdraft is right within the cloud mass itself. If this downdraft is sufficiently compact, the downward velocities can be quite fast.

There are three primary processes that act to produce these downdrafts. One is called precipitation loading. As precipitation falls through the air, it drags the air along with it, causing a downdraft. Almost all precipitation produces at least some downward air motion from precipitation loading.

A second cause of the downward motion is from evaporation. As precipitation falls into unsaturated air (e.g., below cloud base), evaporation of the rain begins. It takes energy to evaporate water, and that energy must come from the surrounding air. If energy is removed from the air, the air must cool. Cool air is heavier than warm air at the same altitude, so it tends to fall toward the ground just as water (which is heavier than air) falls when poured from a bucket. The farther it has to fall and the more evaporation that takes place, the faster the downdraft will be.

A third cause of some downward velocity is the melting of ice. As previously mentioned, much precipitation in the atmosphere forms as ice high in the clouds. As this frozen precipitation falls below the 0°C isotherm, it melts. Again, in order to melt the ice, it takes energy from the surrounding air. The net result of this is to cool the air and cause the cold air to accelerate downward.

When one or all of these processes are working, the result can be a strong downdraft, often of moderately small size. As the downdraft approaches the surface of the earth, it spreads out. This is frequently compared to the process of spraying water from a hose onto the ground. When the water hits, it cannot go into the ground but must spread out horizontally. If the hose is aimed straight down, the water will spread out in all directions at equal speed. If the hose is aimed at an angle to the ground, the water will move faster in one direction than in another.

Real microbursts act in exactly the same way. Those that come down nearly vertically produce winds that radiate out from the center with nearly equal winds in all directions. If they come down at an angle or, more commonly, if they come from a storm which is itself moving horizontally with some speed, the wind speeds

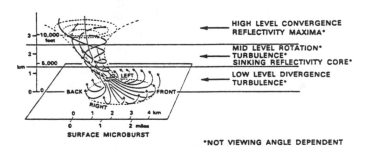

Figure 9.2 Schematic illustration of the major components of a microburst. After Fujita, 1985.

on one side of the microburst will be much faster than on the opposite side. If the storm movement is fast enough, almost all of the wind from the microburst will be in the direction of storm movement with little or no wind moving backward. Figure 9.2 shows schematically the major characteristics of a microburst.

The definition of a microburst given earlier has been modified somewhat to include a measure of the wind speed as well. A working definition of a microburst used in the operational detection of them by radar is that they must have a difference in wind speed (called "shear") from one side of the microburst to the other of 10 m/s or more. If the distance from the strongest winds on one side to the strongest winds on the opposite side is closer together than 4 km, the event is classified as a microburst; if this distance exceeds 4 km, they are sometimes called macrobursts. It is not uncommon that an event that starts out meeting the microburst definition will expand to meet the macroburst definition later on in its lifetime. Quite often the distinction between micro- and macroburst is ignored and the whole event is called a microburst.

What causes a microburst to be detectable by radar? And, what does a microburst look like on a radar? As far as the source of the echo is concerned, we need to differentiate between two kinds of microbursts, wet and dry. A wet microburst is one which occurs with appreciable amounts of rainfall near or coincident with the microburst. In this case, the microburst is detectable because of the presence of the raindrops. While the raindrops are falling relative to the earth, their horizontal motions are determined by the horizontal winds generated by the microburst. Wet microbursts are quite easily seen by most Doppler weather radars.

Dry microbursts, on the other hand, do not have precipitation falling directly in the microburst;

sometimes there is no precipitation falling at all. Microbursts do, however, always require a cloud to produce them. But what happens in some microbursts is that the cloud forms, and precipitation-sized particles form and begin to fall from the cloud. Through precipitation loading, evaporation and sometimes melting, a downdraft is formed. The precipitation evaporates completely some distance below cloud base, but the downdraft continues downward, producing a microburst at the ground. Virga is the name for wisps or streaks of water or ice particles falling out of a cloud but evaporating before reaching the earth's surface as precipitation; virga is a fairly common phenomena in the dryer parts of the country, especially when cloud bases are high above the surface.

Since there are no hydrometeors for the radar to detect in dry microbursts, if they are to be visible at all, it must be because there is another source of detectable echo. Frequently this echo is caused by the presence of insects in the boundary layer. Microbursts are often strong enough to pick up dust and debris from the surface, and this can also contribute to the detectability of the event.

Now, let us return to the question of what a microburst looks like on the radar display. As mentioned, microbursts are frequently reasonably symmetric about some center point where they contact the ground. Even if this is not the case and the microburst is moving along with the parent storm, it will have velocities which are faster on one side than on the other. A Doppler radar can detect the radial velocity (and only the radial velocity!) of targets moving within its radar beam. No matter where the radar is located relative to the microburst, the radar should see approaching velocities on the side of the microburst closest to the radar and receding velocities on the side farthest from the radar. Except for extremely asymmetric microbursts, microbursts should almost always show the typical microburst "signature" of approaching/receding velocities.

Most modern Doppler radars have displays which show radial velocities as colors. A frequently-used convention is to have the receding velocities displayed in browns and reds with the approaching velocities displayed as greens and blues. This convention was chosen to coincide with the experience of astronomers who long ago observed the red shift: Stars that are moving away from the earth have their light shifted toward the red colors; the expanding universe produces, on average, a red shift. With this color convention in mind, the microburst then has a distinct signature of greens on the near side and reds on the far side of the event. Directly across the middle of the event (perpendicular to the radar beam) where the velocities are neither toward nor away from the radar but are rather tangential or stationary, the radar will display zero velocity.

Color Figure 5 shows a strong microburst detected by the UND radar on 11 July 1988 near Stapelton International Airport, Denver, Colorado. This is obviously an example of a wet microburst; radar reflectivities are as strong as 54 dBZ. This particular microburst occurred close enough to Stapelton that a number of aircraft had to abort their landing approaches.

One of the most significant developments in the detection of microbursts is that there is now a computerized algorithm which will automatically detect microbursts. This algorithm was developed for the FAA by MIT Lincoln Laboratory and is designed to work with both the NEXRAD and TDWR systems. These radars will scan a region around an airport in a particular way and process the data as fast as it is collected. When the radar detects a wind-shear event exceeding certain thresholds (along with some other criteria), a microburst warning would be issued. This message would be relayed to the control tower of the airport, automatically stating the location, intensity, and runways affected by the event. The air traffic controllers would then pass this information along to any aircraft which might be affected. The microburst-detection algorithm underwent real-time tests and evaluations in 1988 at Stapelton Airport in Denver, Colorado, at Kansas City, Missouri, in 1989, and at Orlando, Florida, in 1990. These tests showed this algorithm to be a very valuable technique for detecting potentially deadly microburst events. In fact, this microburst-detection algorithm detected the microburst shown in Color Fig. 5 and issued a warning to the air-traffic controllers working in the control tower at Stapelton.

Gust fronts

Gust fronts are another kind of wind event that are frequently detectable by radar. Gust fronts (also called "thin lines" or "fine lines") form when the outflow from a storm moves away from the storm and into the (usually) clear air ahead of a storm. Strong convective storms very often produce gust fronts that can move well away from the parent storm. Once formed, gust fronts can move great distances, sometimes lasting long after the parent storm has died and disappeared from the radar display.

A gust front is visible because of two features. One is that it will often produce an intensification in the reflectivity field, making it visible on the reflectivity display of the radar. This increased reflectivity comes from the lifting of insects, dust and other surface debris into the atmosphere where the radar beam can intersect it. Occasionally, the return of gust fronts are enhanced by the bird activity that accompanies it. The birds will concentrate in the region of the gust front because there are insects and food available. While the insects are carried along by the gust front, the birds intentionally follow the food supply and move with it of their own volition. The reflectivity of gust fronts is often on the order of 0 to 20 dBZ, occasionally higher.

The second way gust fronts are detected is through their velocity characteristics. The winds on one side of the gust front are essentially the ambient, environmental winds of the pre-storm conditions. Behind the front the wind speed is substantially stronger. In fact, the wind speed behind the gust front must be faster than the speed of movement of the front itself, otherwise there would be nothing to keep pushing the gust front along. Since a Doppler radar can detect the velocities of detectable particles, and if these particles are being carried by the wind as insects, etc., would be, then a Doppler radar can often locate a gust front on the basis of a change in wind speed in the vicinity of the gust front.

One problem with this detection, however, is that a Doppler radar can only detect radial velocity. If a gust front is approaching toward or receding from a radar, it is easily detectable. If it is moving tangentially to the radar beam, there would be little radial component to the motion field and the gust front might not be detectable. This means that a Doppler radar cannot detect all gust fronts based on velocity information alone.

The spectrum width of the velocity field can sometimes be used to indicate the location of a gust front. Spectrum width is a measure of the variability of the wind within a single pulse volume of the radar. At the actual gust front location where storm winds meet environmental winds, there is a lot of variability in the wind field. As a radar scans through this region, it should detect a region of enhanced spectrum width. Indeed, this is often the case during gust fronts. The location of the gust front is clearly seen on the spectrum width display because of the strong variability in velocities that is taking place at that location. This is true even when the gust front is moving tangentially relative to the radar beam.

Color Figures 7 and 8 show a PPI and an RHI through a gust approaching the UND radar at Kansas City on 27 August 1989. This gust front occurred about 30 min before several microbursts were detected in the area. The gust front may, in fact, have been the trigger which helped form the storms which produced these microbursts. [Between 2140 UTC and 0034 UTC on 28 August, there were 21 microbursts detected by the UND radar.] The data for this gust front were collected at 1.5° elevation angle, so it is not as close to the surface as some other tilts might have been. On Color Figure 7 we can clearly see the gust front on the reflectivity data; it is less easily seen on the velocity data. Unfortunately, there is also some second trip echo mixed in along the gust front. On the reflectivity data there are some wedge shaped echoes pointing toward the radar which are not visible on the velocity display; these are the second-trip echoes. Interestingly, the human brain can fairly easily filter these out and see the connection in the reflectivity data which forms the gust front.

In the velocity data the signal is not as easily seen. Near the radar, however, there is some clear-air return which is moving away from the radar, i.e., toward the northwest or north-northwest. At about 15 km range or slightly farther out on the 300° azimuth radial there is evidence of approaching velocities. This band of approaching velocities coincides with the thin line of reflectivity marking the gust front's location. Behind this gust front and in the stronger echo to the north-northwest there are some very strong velocities which are moving toward the radar at a speed of about 27 m/s (i.e., the velocities are aliased).

The RHI data for this gust front are shown in Color Fig. 8 which was taken at an azimuth of 309°, 2 min after the PPI data. The clear-air return in front of this approaching thunderstorm is very shallow and almost undetectable on the velocity display. At the 14-km range of the gust front, we can see a slight increase in reflectivity and height of the clear-air return. In the velocity data there is a small region of greens with mostly orange colors to the left and reds to the right. The point where the oranges meet the greens is where the gust front is; where the greens/blues meet the reds is where the velocity aliases into the wrong colors, indicating velocities approaching at this point of 14.9 m/s (based on a PRF of 1100 Hz and the wavelength of 5.4 cm of the UND radar).

Since we are looking at this figure anyway, notice the region of yellow and red velocities within the main echo of the storm between 30 and 40 km range and 2 to 5 or 6 km altitude. This region is air flowing into the storm (or being "sucked" into the storm as the storm moves toward the radar). This inflow region appears to extend back into the clear air to at least 15 km range from the radar. Once the air enters the storm, it is processed in the main updraft (probably located somewhere near the 35-km range mark). Some of the air that goes up, turns back toward the southeast, and is seen to be flowing toward the radar in the anvil echo (with speeds near the top of the storm on the order of 30 m/s). Some of this air likely also turns downward in the downdraft which is causing the rapid velocity moving toward the radar below 2 km from the gust front all the way back to 38 km.

One problem with this is that it is based on a single vertical profile through the storm. This storm is a three-dimensional animal. Not only is there motion toward or away from the radar (which the radar sees and displays) but there is also motion perpendicular to this plane which the radar cannot see. Thus, the air may be moving back into the storm as described above, but it may also be moving into or out of the figure as well. Similarly, the motions toward the radar in the anvil and at the surface likely have strong components to the right or to the left of the radar beam. It is always fun to try to interpret a storm based on a single view of it, but to do a really good job of interpreting what is happening in a storm requires three-dimensional wind fields. This is difficult to obtain from a single, ground-based Doppler radar. By combining the data from two or more radars, however, it is much easier to get three-dimensional data. The process for doing this is called dual-Doppler processing (or multiple-Doppler processing if three or more Doppler radars are used). This will be discussed somewhat more in a later Chapter.

As far as the importance of gust fronts for aviation is concerned, there are two issues that should be discussed. One is that there is usually not a major hazard involved when an aircraft penetrates a gust front. In fact, it has been said that gust fronts are always "performance-enhancing" events. If an aircraft flies from calm air into the region behind a gust front (i.e., into the gust front), its air speed will only increase. Similarly, if it flies from behind the gust front and into the calm air ahead of the front, its air speed again increases. If it is flying along the gust front, there is no change in its airspeed (but that flight could be a bumpy one!). So, aside from the effects of turbulence, there is no major safety problem.

But gust fronts are important for the aviation industry for another reason. Aircraft generally take off and land into the wind. Gust fronts are often tens of kilometers long and can last for periods well in excess of an hour. Thus, once a gust front passes over a point, the winds at that point will change direction and/or speed rather markedly. And once they change, they usually remain different for quite awhile. So, if an airport knows ahead of time that a gust front is coming, what the winds will be after its passage, and its approximate time of arrival, the runways used for takeoffs and landings can be changed in an orderly way. Without this knowledge, the past experience has occasionally been that all departing and landing aircraft would get lined up to use a certain runway only to have a gust front come through, forcing everyone to taxi to another runway. At major airports this can be a very costly event. By having advanced warning of impending gust frontal passage, the transition from one runway to another can be managed in a much more efficient way.

Meteorologically, gust fronts are important for at least two reasons. One is that the presence of gust fronts indicates that storms are somewhere in the area, and they may be strong storms. Further, gust fronts often help initiate new storms, as was suggested above for the 27-28 August 1989 case. This is especially true where two gust fronts intersect. By watching the progress of gust fronts on radar, it is sometimes possible to anticipate where new storms will develop.

The National Severe Storms Laboratory in collaboration with MIT Lincoln Laboratory developed an automatic gust front detection algorithm for use with radar data (Eilts, 1987). This algorithm was tested at Denver, Kansas City, and Orlando, Florida. It has proven to be a very valuable way to provide airports with useful information on gust fronts for runway management.

Turbulence

The final wind phenomenon that is sometimes seen by radar has already been discussed briefly in the preceding sections. This is turbulence. The spectrum width display of a Doppler radar is a reasonably good indicator of the presence of turbulence, at least in a relative sense. When we look at a display of spectrum width, we often think of it as a display of turbulence. Physically, this is a reasonable thing to do.

However, the radar can show enhanced spectrum width when there is no significant turbulence present. One reason this is possible, for example, is that wind shear between two different layers in the atmosphere can produce strong spectrum widths on radar. As a radar scans the boundary between two such layers, the top of the radar beam will be sampling one velocity while the bottom of the beam will be sampling another. The combination of velocities within a single sample volume is interpreted by the radar (and correctly so) as having a high spectrum width, yet no turbulence may be present.

Regions of stronger spectrum width are frequently seen in several places. As mentioned in the last section, gust fronts and the boundaries around microbursts typically have enhanced regions of spectrum width. Another location where there are often stronger spectrum widths is on the boundary of thunderstorms. And as just discussed, at the boundary between layers of different winds, the spectrum width can also be strong.

Perhaps the largest source of erroneous spectrum width, however, is caused by velocity aliasing. Wherever velocities are fast enough that aliasing can occur, there will always be some radar sample volumes that contain velocities just under and just over one of the limiting velocities. Then a single sample volume will contain a very large spectrum width. Consequently, wherever a region of aliased velocities exists, there will usually be a boundary of high spectrum width surrounding it. This artificially generated source of spectrum width makes spectrum width generally less useful in operational settings.

The ability of weather radar to detect regions of turbulence in the atmosphere is, unfortunately, limited. True clear-air turbulence is usually not detectable by normal weather radars. If turbulence is detected in or near strong thunderstorms, that is probably not particularly important because aircraft should be avoiding these locations anyway. If turbulence is detected around microbursts, there is little reason for an aircraft to be flying there because of the other more serious hazards that exist from the microburst itself. So, the only place where turbulence detection might be useful is in weak to moderate, widespread precipitation echo. Turbulence in these regions is not a serious problem.

Clutter

One final source of "clear air" return must be mentioned: clutter. Actually, clutter is a misnomer applied to the echo from ground targets by people who are interested in detecting weather echo (i.e., by many meteorologists). However, one man's signal is another man's noise. If you are truly interested in detecting meteorological echoes, ground targets may be a very obnoxious form of clutter.

But sometimes ground targets can be your friend. For one thing, ground clutter is always there. It is always reassuring to turn on the radar and see a familiar pattern of the nearby ground pattern. This indicates the radar is working and may even indicate it is working correctly. Radars have been known to display targets at the wrong azimuths. If you recognize your local clutter pattern, this should be immediately obvious to you. You cannot use weather targets this way because they keep moving around!

Ground targets are also useful for checking the range of your system. It is always nice to have a good radio tower or two that you know exactly where they are (see Chapter 13 for techniques to determine this). By measuring the range to these targets (carefully), you can be assured that range is correct.

And ground targets can be used as quantitative checks on the overall system sensitivity. From the radar equation for point targets, for example, the power received by a radar is a function of the transmitted power, antenna gain, range and target strength (backscattering cross-sectional area σ). A radio tower or other good target should always return about the same amount of power to the radar (albeit, it is usually displayed on a PPI or RHI as radar reflectivity factor Z

instead of received power Pr or σ). If you check your favorite tower and notice that the displayed signal is weaker than usual, either your transmitter or receiver probably has a problem. Of course, the tower might have changed or your antenna gain might have changed, but these are less likely (unless the wind damaged the antenna or tower or both). Be aware, too, that additional antennas are occasionally put on existing radio towers, so their backscattering cross-sectional areas are not always constant.

Ground targets can even be used to recover miscalibrated radar data. For example, the NCAR CP2 radar operated all of its 1972 and 1973 field seasons with an incorrect receiver calibration, producing errors in the reflectivity data of 11 to 12 dB. Because the field seasons had already passed into history before this was fully realized, it was not possible to go back and recalibrate the system, but it was possible to utilize the constancy of the nearby ground target's backscattering cross-sectional areas to recover the magnitude of the calibration error. This was done by identifying several hundred targets that showed up at the same locations for both 1972 and 1974 and determining the average difference in their signal strengths between these two years. This difference was then applied to the 1972 radar reflectivity values, producing reflectivities which were much closer to correct than they were before (Rinehart, 1978).

Color Figure 1 shows an excellent example of ground echoes (see Color Fig. 6 also). Near the radar are the local hills, buildings, trees, etc.; 30 to 50 km to the north are some additional hills and echoes; 80 to 120 km to the west are the Rocky Mountains. By comparing the ground pattern of the Rocky Mountain area with maps of the area, it is easily possible to locate specific geographic locations. Pikes Peak, for example, is located about 115 km from the radar at 220° to 215° azimuth. There are also some useful radio towers on this display, but they are typically quite small. One is located at a range of 48 km and an azimuth of 296.5°. If you really want to locate radio towers, however, it is much better to magnify the display considerably so they are more easily visible.

Don't confuse moving point targets with ground targets. There are some moderately strong point targets located in the clear region toward the west southwest which are not ground targets: They have velocities! These are either individual birds, or, more likely, low flying aircraft (e.g., some may be approaching or leaving Stapelton International Airport at Denver).

In conclusion, ground clutter is often not clutter at all. It can serve very useful purposes on occasion. To sanitize a data set by filtering out all of the ground targets may make the data set "look better" and may make automatic algorithms operate better, but it does eliminate a source of echo that can be of value. On the whole, I suppose, it is better to get rid of it, but do so grudgingly. Ground targets are some of my favorite targets!

Chapter 10

Advanced Uses of Meteorological Radar

Until now we have concentrated on the basic theory and uses of radar in meteorology. Almost all of the uses mentioned so far have been extended far beyond the simplistic descriptions given. Numerous journal and conference papers attest to the wide variety of uses to which researchers have applied radar. In the following sections we will explore a number of these in slightly more detail.

Rainfall measurements

One of the earliest quantitative uses of meteorological radar data was for the measurement of rainfall. Radar's ability to scan rain showers and thunderstorms over large areas very quickly made it obvious to the early users that much could be learned about rainfall through the use of radar.

As we have seen, the power returned to a radar is dependent upon the radar reflectivity factor Z of a weather echo. Reflectivity, in turn, is dependent upon the number and diameters (raised to the sixth power) of the raindrops present in the radar's sample volume.

Chapter 10

Rainrate is also dependent upon the number and diameter (raised to the third power) of the raindrops and also upon their terminal velocities.

The relationship between reflectivity and rainrate has been studied extensively since the early days of radar meteorology. As mentioned, many empirical Z-R relationships have been developed and used to convert radar measurements into rainfall estimates. Natural variability in the drop size distributions within an individual storm, however, makes the simple use of Z-R relationships less than perfect for measuring rainfall. It was not unusual to have radar estimates of rainfall and raingage measurements made at ground level to differ by a factor of two on some occasions (i.e., 100% error).

One of the perennial questions in any comparison between radar and raingage estimates of rainfall is which is correct. Most researchers, especially the earlier ones whose backgrounds were in hydrology, tended to believe the raingages were correct and used them as ground "truth." However, radars and raingages really measure different things. The sample volume of a raingage is very small compared to that of a radar. Typical raingages have openings on the order of 8 to 12 inches across (0.03 to 0.07 m^2 sampling area). The typical area covered by a single pulse of a radar is 150 m long (corresponding to a pulse duration of 1 μs) and perhaps 1000 m across (for a 1° beamwidth at a range of 57 km from the radar). Thus, the sample area of a radar is more than 2 million times larger than the sample area of a raingage.

This points out one of the major advantages of using radar for measuring rainfall. In order to make good measurements of rainfall over an area using raingages, it is necessary to have a fairly large number of gages. The actual number depends upon the kind of rain being sampled as well as the accuracy desired in total rainfall. For widespread, relatively uniform rainfall, a single raingage might be adequate. However, most rainfall is variable; rain from thundershowers is extremely variable. In order to measure the rainfall from a thundershower with an accuracy of, say, 10%, it might be necessary to have several hundred raingages over the area covered by the storm. In many parts of the country much of the annual rainfall comes with only a few, heavy storms. Thus, if we really want to get accurate rainfall measurements over much of the country, we might need millions of raingages spread around. Radar's ability to see large areas nearly simultaneously is one of its major advantages.

Another source of difference in rainfall estimates of the two different instruments is that radar usually looks somewhere above the surface of the earth while a raingage is located right at the earth's surface. It takes time for rain measured by a radar to fall to the earth. A sample of rain which might be above the raingage when the radar measures it can move some distance from the raingage by the time it reaches the earth's surface. This is an especially significant problem in fast moving, showery rainfall where the rainrate can change dramatically from one location to another. Evaporation between the radar sample and the ground can also cause changes in rain before it reaches the surface.

Another problem in measuring rainfall is beam blockage. Radars used for measuring rainfall sometimes don't have an unobstructed view of the area they need to see. If that happens, it is sometimes possible to correct for the blockage by increasing the measured value of power received by an amount that is related to the fraction of the beam blocked. Figure 10.1 shows the blockage of a beam by obstructions on the horizon. If measurements of this blockage are made in the direction of interest, it is possible to adjust measured values to get a better estimate of the rainfall that is actually occurring.

A final source of errors in measuring rainfall with radar is the brightband. During stratiform rain events, precipitation forms as snow, falls through the 0°C isotherm, melts, and becomes rain at the surface, producing a brightband in the process. Earth's curvature can make the beam intersect the brightband at longer distances from the radar. The rainrate calculated using the reflectivity from the brightband will be higher than the actual rainrate occurring at the ground. So, when stratiform rainfall is being detected and conditions are ripe for producing a brightband, be aware that rainfall measurements may be in error. It may even be possible to use knowledge of the physics of the brightband, the height of the freezing level, and the distance of the echo from the radar to correct for this overestimation.

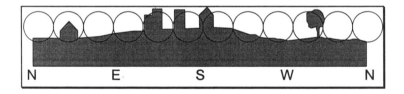

Figure 10.1 Schematic of the beam blockage that might occur for a radar scanning horizontally around a radar. Each circle represents the size of the mainlobe of a radar antenna beam pattern looking in each direction. For real radars, there might be as many as 360 beams, not just the dozen shown here.

There are significant temporal differences between raingage and radar samples. Raingages typically have a temporal resolution of 1 min or longer (some gages are read only daily). Radar, on the other hand, can scan an entire storm in a matter of seconds. To compare these instruments, then, requires that some temporal averaging be done to both sets of data.

Rainfall measurements by radar have continued to be an important use of radar data in recent years. A number of schemes have been used to improve the rainfall estimates from radar. One of the simplest, at least conceptually, is to combine raingages and radar data together. Raingages are used to make accurate rainfall measurements at specific points within the radar field of view. These point measurements are then used to "calibrate" or adjust the radar data at those points so that both systems give the same values. These correction factors are then applied to the radar data around these points to improve the radar rainfall measurements. When this kind of processing is done, the radar/raingage combination can give rainfall measurements which are within about 10 to 15% of what a large raingage network might provide for the same storm. It is now possible to have raingages located in a radar's field of view which automatically record the rain falling and transmit the data via telephone line, radio or satellite to the radar location where it can be incorporated into the radar data in real-time. This is very useful information for hydrologists and flood forecasters for operational uses.

Dual-wavelength radar

One of the more specialized uses of radar, at least on a research basis, is to use two radars of different wavelengths to view the same volume in space to detect the presence of hail within storms. There are a few special purpose, dual-wavelength radars available which can make these measurements. The Russians, in fact, have used dual-wavelength radars fairly routinely for hail detection on some of their hail-suppression, weather-modification projects. Much research has been done on this topic since it was first proposed in the early 1960's.

Dual-wavelength radars designed for hail detection usually use an S-band and an X-band radar together (i.e., 10-cm and 3-cm wavelengths, respectively). For convenience, the antennas for both radars are usually mounted on the same antenna pedestal, so they point in the same direction as the antenna moves in elevation and azimuth.

The basis for this use was discussed earlier when Rayleigh and Mie scattering were covered. When only raindrops are being illuminated by the radars, all of the drops are within the Rayleigh region. The largest raindrops that can exist in the atmosphere without breaking up into smaller droplets is on the order of 6 or 7 mm. Even at X band, this is moderately small compared to the wavelength. Thus, the radar reflectivity factor z from both radars should be the same, i.e., $z = \Sigma(N_i D_i^6)$.

When hailstones are present, however, the stones can become large enough compared to the wavelength of the X-band radar that Rayleigh scattering no longer applies. By definition, hail has a diameter of 5 mm or larger; if it is less than 5 mm (and ice), it is called ice pellets. In any case, hail and/or ice pellets smaller than 5 mm are of little consequence to crops and property. Most hail damage comes from hailstones much larger than this.

So, when *real* hail is present, the S-band radar will be seeing particles which are in the Rayleigh region (or the low end of the Mie region) while the X-band radar will be seeing particles which are well into the Mie region. The two radars will get different returns from the same sample volume. By comparing the reflectivity from the two radars, the presence of hail is fairly easily detected. In some cases it is possible to estimate the size of the hail as well.

One quantitative parameter which can be derived from dual-wavelength data is called the hail signal (Eccles, 1975). Hail signal is the ratio of the reflectivities at the two different wavelengths, usually expressed on a logarithmic scale. Thus,

$$H = 10 \log_{10} \left(\frac{z_{10}}{z_3} \right) \tag{10.1}$$

where H is the hail signal (dB), and z is the radar reflectivity factor (mm^6/m^3). If only rain is being observed, $H = 0$ dB. If hail is present, H is usually positive, sometimes reaching values as large as $+20$ dB or so. For some very restricted conditions (monotonically distributed hail diameters of certain sizes), it is also possible to have slightly negative hail signals, but nearly all significant hail has hail signals larger than perhaps $+3$ dB or more.

Dual-wavelength radars are not without their problems, however. As mentioned at the start of this section, it is necessary for both radars to sample the same volume in space. This sounds moderately easy. Indeed, early workers in the field of dual-wavelength radar made sure that they had radars with "matched" beamwidths, that they were aimed at the same azimuth and elevation, and that they used the same pulse lengths. Good enough? No!!! As it turns out, the antenna beam *pattern* must also be reasonably matched, including the sidelobes (Rinehart and Tuttle, 1982). While it is possible to choose antennas with the same beamwidth for the mainlobe (e.g., 1° beamwidths were usually used at both S and X band), it is more difficult to get antennas that are a factor of three different in diameter to have the same *sidelobe* pattern. Until this requirement was identified in the early 1980's, the importance of matched antenna patterns was not fully appreciated. Unmatched antenna beam patterns produced hail signal errors as large as 20 dB in some situations. The ideal dual-wavelength radar would be designed to have matched beam widths and sidelobe patterns for both wavelengths. No such radar has ever been built.

There is another use for dual-wavelength radars. This is in the measurement of rain. Rain-measuring dual-wavelength radars have one wavelength which is moderately attenuation-free and another which is affected by attenuation. The unattenuated signal is used to estimate the amount of attenuation in the attenuated wavelength signal. Attenuation is very closely related to rainrate. In fact, at wavelengths near 1 cm, there is almost a one-to-one relationship between attenuation and rainrate. Rain-measuring dual-wavelength radars sometimes use 5-cm wavelength and 3-cm wavelengths.

Polarization diversity

Radars transmit and receive electromagnetic radiation. One of the properties of electromagnetic radiation that has not been examined yet is polarization. In this section we will consider some of the uses of polarization information.

Electromagnetic radiation consists of electric and magnetic fields which oscillate with the frequency of radiation. These fields are always perpendicular to each other. When electromagnetic radiation leaves a simple wire antenna, the magnetic field is perpendicular to the radiating element while the electric field is in the same plane as the element. Since the electric and magnetic fields are always perpendicular, we can specify the orientation of the electromagnetic radiation by specifying the orientation of either of these fields. Convention has it that we specify the orientation or "polarization" of an electromagnetic field as being the orientation of the electric field.

The polarization transmitted and received by a radar can be chosen to perform certain tasks. Many meteorological radars have antennas that transmit and receive horizontal radiation. Since raindrops tend to flatten out somewhat as they fall, they are wider horizontally than vertically. If you want to maximize the rainfall detection capabilities of a radar, you might choose to use a radar with horizontal polarization.

Those interested in detecting aircraft flying within a rainstorm, however, consider rainfall as clutter and attempt to minimize its return while at the same time maximizing the signal received from the aircraft. By using circular polarization (in which the orientation of the electric field rotates with time along the direction of the propagation path), the return from nearly spherical raindrops can almost be eliminated while that from nonspherical targets such as aircraft are still clearly detected.

There are a number of ways polarization can be used to increase our knowledge of meteorological targets. Polarization techniques have been used to determine the sphericity of raindrops, the orientation of ice crystals in the atmosphere, the presence and characteristics of hail, and to improve rainfall

Chapter 10

measurements. Let us explore some specific items as examples.

Circular depolarization ratio

One of the parameters available through processing of polarization data is the circular depolarization ratio (CDR) of the received signal from a dual polarization radar. Radars capable of detecting CDR transmit right-hand circular (RHC) polarization and receive both right- and left-hand circular (LHC) polarizations using two separate antennas. CDR is defined as the ratio of the equivalent radar reflectivity received by the so-called parallel component (i.e., the same polarization as that transmitted) to the equivalent radar reflectivity received by the so-called orthogonal component, usually expressed in decibels, i.e.,

$$CDR = 10 \, log_{10}\left(\frac{z_{\parallel}}{z_{\perp}}\right) \qquad (10.2)$$

$z_{\parallel}$ and $z_{\perp}$ are the reflectivity in the parallel channel and orthogonal channels, respectively. By appropriately processing these received signals, it is possible to estimate the degree to which the detected hydrometeors are nonspherical. In principle, CDR can vary from 0 dB to -∞ dB. Infinitely long and thin scatterers would give $CDR = 0$ while perfect spheres would give $CDR = $ -∞ dB. CDR was especially useful in distinguishing hail of damaging size from small hail and rain (Barge, 1970).

CDR has also been used to study lightning discharges in thunderstorms by watching for changes in CDR (McCormick and Hendry, 1970). In thunderstorms, some ice crystals can actually be forced into a common orientation direction because of the strong electric fields that exist. When this happens, the polarization effects are such that this common orientation can be determined. When a lightning discharge takes place, the electric field is suddenly dissipated, and the ice crystals are able to assume more random orientations. The electric field will again build up until the next discharge takes place, repeating the cycle with each lightning stroke.

Linear depolarization ratio

Another way to learn about what is going on inside storms is to use two linear polarizations. This is very similar to the use of circular polarization except that horizontal and vertical linear polarizations are used. Again, two antennas are used. One is used to transmit and receive horizontal polarization while a second antenna is used to receive vertical polarization. The reflectivities are processed similar to CDR, but the resulting parameter is called the linear depolarization ratio LDR, defined as follows:

$$LDR = 10 \, log_{10}\left(\frac{z_H}{z_V}\right) \qquad (10.3)$$

where z_H and z_V are the horizontal and vertical reflectivities, respectively. For perfect spheres, LDR approaches negative infinity (in decibels); antenna limitations with real radars limit LDR values to -40 dB or so for small spheres. For long, thin targets (such as chaff), LDR approaches 0 dB. Real meteorological targets give LDR values typically between -15 to -35 dB. One use of LDR is to distinguish between rain snow and melting snow. In a brightband situation, for example, above the bright band LDR will be one the order of -25 dB, in the melting layer it might be -15 dB and in the rain below, LDR would be -30 dB or less.

Reflectivity depolarization ratio

Another polarization parameter that has proved useful over the past couple of decades is called the reflectivity depolarization ratio (Z_{DR}; also written as ZDR) (Seliga and Bringi, 1976). In this scheme a radar will transmit a pulse of horizontally polarized radiation and receive and process the received echo. On the next pulse it transmits and receives a vertically polarized signal. It continues doing this on alternate pulses, storing and averaging the horizontal and vertical polarization signals separately. Figure 10.2 shows the various components in a radar used to measure Z_{DR}. To use this system, the radar must have a switch which is capable of changing the polarization very rapidly. Most dual-polarization radars of this type transmit with a PRF on the order of 1000 Hz, so the switch must be capable of operating at this rate. And since it is switching a transmitted signal as large as a megawatt, it takes a switch capable of handling high power.

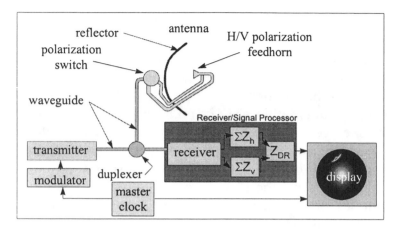

Figure 10.2 Block diagram of a radar to measure Z_{DR}.

The signals that are received from each polarization channel are averaged separately, and radar reflectivity factors are determined from each, giving z_h and z_v. The reflectivity depolarization ratio is defined as

$$Z_{DR} = 10 \, log_{10}\left(\frac{z_h}{z_v}\right) \qquad (10.2)$$

and is measured in decibels.

The shape of raindrops falling in the atmosphere varies from nearly perfect spheres for small droplets up to a couple of millimeters in diameter to more flattened drops up to 5 or 6 mm across. These flattened drops give stronger returns at horizontal polarization than at vertical. Thus, Z_{DR} varies from near zero for spherical droplets to values as large as +5 dB for echoes from large water drops. This added information is useful for refining rainfall measurements made by radar.

Z_{DR} is also useful for indicating the presence of hail. When there are strong reflectivities present, Z_{DR} from raindrops should get moderately large. However, when hail is present, Z_{DR} often goes to near zero. The reason for this is that hail stones do not fall with a preferred orientation as do raindrops. Instead, hailstones generally tumble as they fall. In a single pulse volume of a radar, hailstones will have random orientations. Thus, even if the hail is not spherical (the usual situation), there will be roughly as many hailstones oriented vertically as horizontally so that polarization effects cancel.

One final exception to some of this is the detection of graupel by dual-polarization radars. Graupel is small, conically shaped hail. The conical shape persists because the graupel particles *do* fall with a preferred orientation. They fall with their flat sides aimed downward and their conical ends aimed upward (at least on average). Once they grow beyond a certain point, however, they, too, will begin to tumble and loose their conical shapes. But while falling as conical graupel, they do have a common orientation. Further, the vertical size of graupel can be larger than the horizontal dimension. This results in slightly negative Z_{DR} values. So, it is possible to detect the presence of conically shaped graupel in some convective storms using Z_{DR} measurements.

Just as with dual-wavelength radars, there are antenna mismatch problems with dual-polarization radars (Herzegh and Carbone, 1984). Even though a Z_{DR} radar, for example, uses exactly the same antenna for its horizontal and vertical signals, the details of its antenna beam patterns will not be the same for both polarizations. If the differences are big enough and the reflectivities of the storms being detected variable enough, significant errors can occur in dual-polarization measurements. No measurement system is perfect. If we understand the limitations of a system, however, we can still get useful measurements from it.

Dual-Doppler processing

A single Doppler radar can measure the radial component of a target but cannot give its complete three-dimensional velocity without making some assumptions, adding some additional information, or doing some serious guessing. In principle, however, if we can measure the radial velocity of a target from three different locations, we could determine its true three-dimensional velocity. Over the years a number of schemes have been developed to combine the velocity data from two or more Doppler radars to produce two- and three-dimensional wind fields from a variety of storm situations.

Dual-Doppler processing has been particularly useful because it requires only two radars, not three. Unfortunately, the requirement for having three different components of motion to get three-dimensional winds cannot be eliminated. With two radars, it is only possible to directly determine two components of the velocity field. This is sometimes all that is needed, however. For example, in studies of the asymmetry of microbursts, two Doppler radars can scan horizontally and determine the horizontal velocity of the microburst's wind field. MIT Lincoln Laboratory and the University of North Dakota have operated a pair of radars in several locations between 1985 and 1993 for exactly this purpose. In fact, over 15,000 dual-Doppler scan pairs were collected over a 3-yr period (Bob Hallowell, personal communication)!

Chapter 10

It is possible to combine radial velocity data from only two Doppler radars and get three-dimensional wind fields if you include some other assumptions in the processing. There are actually four unknowns that we need to solve for, so we need four equations. The unknowns are u, v, w, and w_t where these are the components of velocity in the x, y, and z directions and wt is the terminal velocity of the precipitation. The two Doppler radars provide two velocity estimates. To supplement these two measurements, the usual technique is to use the mass continuity equation and to assume that the terminal velocity is a function of the radar reflectivity. This latter relationship is based on a Marshall-Palmer drop size distribution of raindrops and the laboratory measurements by Gunn and Kinzer (1948) of terminal velocity as a function of drop diameter. By adding these assumptions to the radar measurements, it is then possible to calculate the horizontal and vertical components of the wind at every point (provided the radar data from both radars extends from the surface to the top of the storm). A boundary condition is also required or typically added, namely, that the vertical component of the wind is zero at the bottom of the storm (since the earth's surface is not porous, this is a reasonable assumption if the data start close enough to the surface). Sometimes it is also assumed that the vertical component of the wind is zero at the top of the storm, but this is frequently not true, especially for growing or dying storms.

When three radars are available, it is possible to calculate the complete three-dimensional wind without resorting to the assumptions mentioned above. It may still be desirable to constrain the calculations by making some of these assumptions, but it is not theoretically necessary. If four or more Doppler radars are available, there is obviously more data available than is required. However, in this "over-determined" case, the additional data are still valuable as it can be used to reduce errors which might otherwise exist.

To be really useful, the data from a dual- or multiple-Doppler network must be collected from different points of view. That is, if two of the radars are looking at the storm from the same direction (i.e., if the storm is on the baseline connecting the two radars), they will both measure the same component of velocity; no more information will be available than if there were only one radar present. Thus, the radars need to be viewing the storm from different directions. The ideal situation would have the two radars exactly 90° apart. Unfortunately, the beams of two radars can only intersect at 90° along a circle[10] with the two radars across from each other (a diameter apart). Measurements along a single line would not provide much information. Consequently, we are required to accept data from points where the intersection angle between the beams is more or less than 90°. Figure 10.1 shows the locations of useful dual-Doppler lobes for two radars. When three or more radars are operated as a multiple-Doppler network, the useful area is a relatively small area which is between all of the radars (unless the radars are operated as a network of *dual-Doppler pairs*, in which case the radars are not really being used to improve the quality of the measurements but instead the areal coverage).

An example of the results of doing dual-Doppler processing is shown in Fig. 10.2. The data for this

---

[10] Actually, the beams will intersect at a 90° angle along the inside of a sphere; for horizontal measurements, the locus is a circle.

analysis came from the Lincoln Laboratory FL2 S-band radar and the UND C-band radar on 9 August 1985 at 144035 CST while they were operating near Memphis, Tennessee. The data are given relative to the location of the FL2 radar (which is located at 0, 0 in the Cartesian coordinates along the edge of the figure); the UND radar was located 1.2 km south and 14.6 km west of FL2. Figure 10.2 shows both reflectivity and wind information. The reflectivity data are given as solid contours starting at 0 dBZ on the outside of the echo and increasing to more than 40 dBZ at the inner-most contour surrounding the microburst; the peak reflectivity at this time was near or slightly above 50 dBZ (but averaging and processing of the data tend to reduce the peak values recorded by either of the radars). The wind data are shown as many small vectors which give the wind direction and speed. Direction is given by the orientation of the vectors (the arrowheads point in the direction the wind is blowing). Speed is given by the length of each vector. The vectors are spaced at 200-m intervals in both the x and y directions. A vector 1 km long represents a wind of 25 m/s.

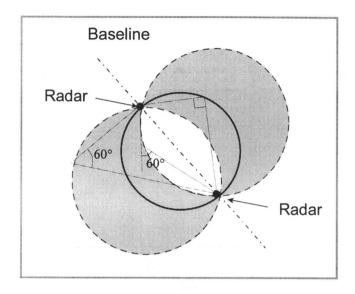

Figure 10.1 Schematic of the location of dual-Doppler lobes for a pair of Doppler radars. Useful data would be limited to the shaded area within the larger circles (except for the area within these circles which is along the baseline). The radar beam intersects at 90° along the central circle. The beams intersect at 60° along the dashed circles. Useful data can sometimes be obtained by enlarging the circles so the beams intersect at 30°.

The data in Fig. 10.2 are from a microburst occurring under a moderately strong thunderstorm. This is obviously a wet microburst because the reflectivities are so high.

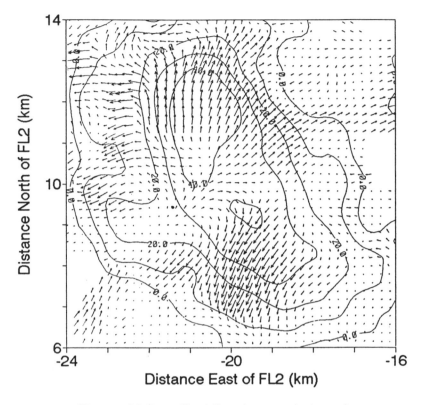

Figure 10.2 Dual-Doppler analysis of a microburst near Memphis, Tennessee, at 144035 CST on 9 August 1985. The data for this came from the Lincoln Laboratory FL2 and the UND radars.

NEXRAD and TDWR

Next Generation Radar (NEXRAD, WSR-88D)

Radars, just like any other living organism, are constantly undergoing change and evolution. Meteorological radars, too, have changed considerably over the years. What exists now as state-of-the-art will someday be relegated to a museum or a history text. So while we may be enamored with the latest and greatest technology, we must realize that "this, too, shall pass" and become obsolete. Consequently, any discussion of the current "flavor" of meteorological radar is subject to change in the near future. This evolution will never end. As long as we keep having new requirements, new computer capabilities, new hardware capabilities – and, most of all, new ideas from creative new minds - radar will continually change.

Originally, radars were mostly designed to detect military targets. By the early 1950's radars were designed specifically for weather detection. The AN/CPS-9 X-band radar was the first military radar designed specifically for meteorological use.

The first full-fledged radar for the Weather Bureau (as it was called then) in the United States was

Chapter 11

the WSR-57 S-band radar (Weather Surveillance Radar, commissioned in 1957). At the time of its introduction the WSR-57 was a tremendously capable machine. It served mankind very well for over three decades. But it was clear after twenty years of use that it's useful life was limited and it would have to be replaced.

In 1979 the Joint Doppler Operational Project of the office of the Federal Coordinator for Meteorological Services and Supporting Research conducted an investigation into the use of Doppler radar for the identification of weather hazards. The conclusion of this study was that Doppler radars offered significantly improved detection of various meteorological phenomena and that Doppler radars should become a part of the tools available for operational meteorology in this country. In 1982 the federal government initiated the acquisition of a new breed of radars for use by the National Weather Service, the Federal Aviation Administration, and the Air Weather Service and Naval Oceanography Command in the Department of Defense. This Next Generation Radar (NEXRAD) system was to use Doppler capability and would in general upgrade the nation's weather radar system to a state-of-the-art system capable of automatically monitoring and recognizing certain weather events.

The NEXRAD system went through a lengthy and thorough procurement procedure which included an open competition by manufacturers for the design of a new radar system. After initially selecting three groups to submit more detailed designs, two manufacturers were chosen to build prototype radars. One of these was finally selected as the winning design. Once the selection procedure was completed, the NEXRAD system was formally designated WSR-88D, which stands for Weather Surveillance Radar, commissioned in

1988, and having Doppler capability. In the following discussion, both terms will be used for the same system. The specifications of the WSR-88D are given in Appendix D.

More than 150 WSR-88D radars were manufactured. They were installed mostly in the conterminous United States but they have also been installed in Alaska, Hawaii, Guam, Korea, Okinawa, the Azores and in Puerto Rico. The radar is supporting the missions of the National Weather Service (NWS), the Federal Aviation Administration (FAA), and the Department of Defense.

An individual WSR-88D unit consists of three subsystems: the radar data acquisition subsystem (RDA), the radar product generation subsystem (RPG) and the principal user processing (PUP) subsystem. The radar data acquisition subsystem contains not only the hardware usually associated with a radar (e.g., transmitter, receiver, antenna, tower, etc.) but also the software and computer necessary to produce the reflectivity, velocity and spectrum-width data sets and prepare them for high-speed transmission to other locations. Before transmission elsewhere, the data will have already been decluttered, dealiased and range unfolded. "Decluttered" means that the data will have had ground clutter removed; WSR-88Ds have clutter filters which will remove clutter signals up to 50 dB above noise level; clutter maps can also be used to remove more of any residual clutter that remains. Velocity dealiasing is also performed automatically to eliminate velocity ambiguities in the meteorological data. And finally, the data are automatically put into the correct range interval. The intention is to produce a clean data set that contains only meteorological data and that these data be as correct as possible before going on to the next steps in their use.

Once clean, quality-checked data are available for further use, computer algorithms will process the data to produce various products. Table 11.1 lists some of the basic products along with other, more advanced products and algorithm outputs. For a more complete description of these algorithms, see JSPO, 1985 and JSPO, 1986.

Table 11.1 List of selected NEXRAD products
Reflectivity base products
Composite reflectivity
Layer reflectivity
Echo tops
Mean radial velocity
Storm relative mean radial velocity
Radial and azimuthal shear
Combined shear product
Spectrum width base product
Layer turbulence
Severe weather analysis display
Weak echo region
Cross section products
Radar coded message
Combined moment
Vertically integrated liquid water and severe
    weather probability products
Storm track information
Storm structure
Hail index
Mesocyclone and tornado vortex signature
    products
Surface rainfall accumulation (1-h, 3-h, & storm
    total)

VAD winds
Transverse wind*
Microburst algorithm
Gust front algorithm
Icing algorithm

*One of my personal favorites (Hamidi et al., 1979).

Some of the above products will be available all of the time (e.g., the base products) while others will be available only some of the time and/or at only certain locations. For example, microbursts and gust fronts are of interest primarily in the vicinity of airports. WSR-88D radars located far from major airports would probably not run these algorithms. The icing algorithm would not be run during summer months anywhere and might only be used during winter months at some WSR-88D sites.

One of the design features of the NEXRAD system is to make it possible to add new algorithms as they are developed and as new needs arise. Computer processing of the radar data makes this possibility much more likely than if the products had to be produced by hardware alone. The WSR-88D system is currently undergoing a series of revisions which will make the system easier to change and more open to new advances in computer systems. The original system was quite computer-specific; the new system will allow adoption of faster and more powerful computer systems. This should make the radar system much more flexible and allow it to evolve in a more graceful and gradual way.

Chapter 11

Figure 11.1 shows the antenna tower and radome for the second operational NEXRAD radar located at Melbourne, Florida.

Figure 11.1 WSR-88D radar tower and radome for the Melbourne, Florida, radar site. This is the second operational radar in the NEXRAD network.

NEXRAD and TDWR

So, what does the WSR-88D detect and what do its products look like? The color figures section contains a number of examples of weather and nonweather situations.

Color Figure 15 is an example of the storm track algorithm output. It shows the current position of storms on a WSR-88D radar display (marked by the colored reflectivity data and the open circle at the center of each main storm cell) along with the past position of each cell at 15-min intervals (solid black dots) and the forecast position of the same cells at 15-min intervals (plus signs). From this display it is possible to forecast the movement of existing storms over the next hour.

Color Figure 16 shows an example of the storm total precipitation algorithm output. This storm and the area covered by this display is the same as shown in Color Fig. 15 (except that Color Fig. 16 includes data from 0533 UTC through 2204 UTC on 21 March 1991 whereas Color Fig. 15 is for a time of 2108 UTC on 21 March). The color bars on this figure show that the individual storm cells on 21 March produced as much as 4 inches of rainfall in some locations. Note that some of the apparent storm tracks cross. This likely occurred because storms at one part of the time period moved from one direction while storms at another time moved in somewhat different directions.

Color Figs. 17 and 18 are from the Mayville, North Dakota, WSR-88D radar and show a number of summer thunderstorms that were moving from Canada across North Dakota and into Minnesota. Again, the storm-tracking algorithm has marked the past locations and predicted locations at 15-min intervals on the reflectivity plot (Fig. 17). Figure 18 is the Doppler velocity field for the same storms. The general direction of movement can easily be seen by the approaching (green) and receding (red) regions. But within specific echoes there is even more information available. A careful examination of these velocities can reveal subtle yet important regions of convergence, rotation and, occasionally, microburst activity. Unfortunately, WSR-88D velocity data are probably underutilized. It is quite common for TV weathercasters to brag about the new "Doppler radar" but never show the Doppler data. [Sorry! But that's another one of my pet peeves!]

Color Fig. 19 shows a hurricane detected by the Mobile, Alabama, WSR-88D as it slowly made its way on shore on 19 July 1997. Hurricane Danny was not a particularly strong storm as measured by wind speed, but its very slow movement allowed it to dump over 20 in of precipitation on the Gulf coast as it moved inland. The rainfall, measured by radar(!) caused flooding in low-lying regions.

And now, for something completely different, consider Color Fig. 20. This is an image from the Nashville, Tennessee, WSR-88D collected at 1108 UTC on 12 August 1995. It shows two general kinds of echo. Near the radar is an extensive, widespread region of weak echo. Farther out are at least five "donut" echoes. So, what is causing these? [I'm trying to keep you in suspense just a little!] The nearby echo is likely an example of clear-air return caused by insects. The donuts, however, are birds!

How do we know they are birds? Trust me. [I tell my students that a good scientist never trusts anyone, including themselves!] In 1986 the MIT Lincoln Laboratory FL2 radar detected exactly the same kind of ring echoes while it was operating just west of Huntsville, Alabama. Having seen pictures of similar echoes from the 1950's (Elder, 1957), birds were immediately considered a possibility. The center of this echo was over Finlay Island in Wheeler Reservoir west of Huntsville about 25 km. I drove out one morning an hour before sunrise and, as the sun came up, I could see thousands of starlings taking off for their daily feeding. Notice on Fig. 20 that the time is 1108 UTC; this is 508 CDT; i.e., just after sunrise!'

Based on the guesstimated size of a single starling, that they looked to the radar like spheres of water, that they were small in diameter compared to the wavelength, and that they produced a beam filling target, I estimated that there were more than 700,000 starlings roosting on the Island every night. Interestinlgy, the southern-most donut echo on Fig. 20 is located over Findlay Island, yet the picture is 9 yr after the detection of birds at that location by the FL2 radar.

There is one other kind of "donut" echo detected by WSR-88D (and other) radars. This is one where the radar is at the center of the donut. These, too, are often caused by birds, but in this case they are migratory birds. It has been frequently observed by NEXRAD radar operators that rings of weak echo sometimes surround a radar site from perhaps an hour after sunset until an hour before sunrise, especially in the spring and fall. Songbirds (paserines) migrate at night. These small birds take off after sunset, climb to an altitude assigned by their internal air traffic controllers (probably based on temperature and almost certainly based on having favorable tail winds), and fly until near sunrise when the land and feed during the day. Depending upon the vertical distribution of the birds, they can produce distinct rings around a radar with weak or no echo right at the radar followed by a circle of nearly uniform reflectivity echo. These rings are observed mostly on nights when the winds are favorable. If there are strong headwinds present, the birds simply wait until better conditions.

All of the images above were essentially just what the radar sees (along with tracks of where it has been and where it is expected to go). NEXRAD computers can also process the data to produce additional products. Color Fig. 16 is an example of estimating rainfall from a sequence of reflectivity samples. As another example of what NEXRAD can do with velocity data from the WSR-88D, consider Fig. 11.2. This figure is what is called a VAD Wind Profile. It shows the winds above a given radar (again, the Mayville, North Dakota WSR-88D) as a function of time. Time is listed along the bottom of the figure and covers a period of approximately two hours. Height is given along the Y-axis and covers heights from the surface to 50 kft. Each wind barb represents the wind speed and direction at a given time and height. Speed is determined by counting the flags on the barbs. A half flag (*not* at the end of the barb) represents a speed of 5 kt, a full barb is 10 kt, and a triangular flag is 50 kt. Wind direction is given by the direction of the stem of the measurement. The end of the wind barb at the altitude/time location is fixed. The direction of the other end of the barb or stem represents the direction from which the wind is coming. Most of the winds on Fig. 11.2 are from a westerly direction. Most of the speeds are in the 10 to 40 kt region.

Another piece of information on the VAD image is how variable the winds are. On the original image there is color coding to indicate the wind variance. A color scale on the right side (0, 4, 8, 12, 16 kt RMS) gives this information. One use of this is to get an estimate of how reliable the winds really are. If the variance is high, it may be an indication that the winds are not

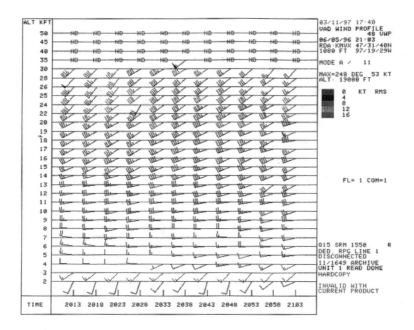

Figure 11.2 Velocity-azimuth display of winds above the Mayville, North Dakota, radar from 2013 until 2103 UTC on 5 June 1996. See text for details on how to read the winds.

quite as reliable as in other locations. Another interpretation of this information, however, is that it is an indication of the presence of turbulence. The higher the variance, the greater the variety of winds in the measurement and, by implication, the greater the turbulence.

One of the limitations of the VAD Wind Profile algorithm is that there must be something present to give a detectable signal. There are two sources of this signal. One is particulates, e.g., rain! The other is Bragg scattering from refractive index variations within a distance that is small compared to the wavelength.

Chapter 11

When rain is the source of the echo, you need to keep in mind that raindrops also have a terminal velocity which may bias the measurement sometimes. And rain does not often exist equally in all directions. Stratiform rain is often shallow but horizontally extensive. Winds from this source of echo might give valid measurements but only over a limited depth. If convective storms are present, they vary tremendously from one horizontal location to another. And they probably don't represent the wind surrounding the shower or thundershower where the echo exists. So winds from these may be badly biased for other reasons. Winds from the clear air are probably most representative of the atmosphere, but these are likely not very deep much of the time, especially during the winter when there is less moisture in the atmosphere to cause refractive index gradients for detection.

As a reminder, there is a network of wind profiler radars being tested in the center of the US to see how well radars can routinely detect winds. These radars are of longer wavelength and are better able to detect clear-air return at all altitudes and all seasons. They, too, suffer from contamination from precipitation. They also suffer from biases introduced when birds are migrating through!

Terminal Doppler Weather Radar (TDWR)

The Federal Aviation Administration recognized part way into the development of the NEXRAD system that not all airports would be adequately covered by NEXRAD. Consequently, they developed specifications for a radar system designed to provide the coverage needed in the vicinity immediately around selected airports. This terminal Doppler weather radar (TDWR) will be used to supplement the NEXRAD network.

TDWR systems operate at C band (5500 MHz) and have a 0.5° beamwidth so that ground clutter can be avoided as much as possible. This is necessary because TDWR systems are operated fairly close to airports, and ground clutter is worst close to the radar. Further, microbursts and gust fronts are usually quite shallow and located near ground level. In order to be detected by radar, the radar must scan as close to the ground as possible. The combination of looking close to the radar and close to the ground makes ground clutter a potentially devastating problem. So, in addition to using sophisticated ground clutter canceling techniques, the use of a narrow antenna beam pattern helps eliminate some of the ground return and makes it more likely that the hazardous wind events will be detected.

The choice of C-band for TDWR was dictated in part by the antenna requirements. A narrow beam width is more easily formed with a short-wavelength radar than with a long wavelength. However, short wavelength radars also produce more velocity aliasing than longer wavelengths (the Doppler dilemma again). Consequently, either velocity aliasing and/or multitrip echo detection is a serious problem with TDWR radars. To help reduce these problems, TDWR radars use an automatic algorithm to scan the storms in the direction of the airport out to long ranges and then select the PRF which minimizes the obscuration of the radar data over the airport. Most of the time this should work reasonably well. On some occasions, however, there will never be a PRF which can give both good velocity coverage as well as unobscured coverage in range. The worst case would be when a long line of storms extending in a line from the radar to the airport and well beyond. Here it might be impossible to pick out real meteorological events from garbaged data due to multitrip echoes and/or range folding. It will always be possible, however, to determine when this is a problem so its occurrence would be recognized.

Chapter 12

Quantification

When you can measure what you are speaking about and express it in numbers, you know something about it; but when you cannot measure it, when you cannot express it in numbers, your knowledge is of a meager, unsatisfactory kind; it may be the beginning of knowledge, but you scarcely, in your thoughts, advanced the state of science.

Lord Kelvin

Radar is capable of providing excellent quantitative data. While we generally simply look at radar displays to see where storms are and about how strong they are, it is possible to get much more information than just superficial qualitative impressions. To gain the ultimate in quantitative information from a radar, however, requires that we make careful measurements of its various components.

The components of a radar which need to be quantified are all of the parameters that go into any form of the radar equation along with auxiliary or housekeeping information such as azimuth, elevation, range, and time. A complete list of parameters which we need to be able to measure includes:

transmitted power (peak and/or average)
transmitted frequency
wavelength
waveguide losses
radome losses
received power
receiver frequency
receiver bandwidth
pulse duration and/or pulse length
antenna gain
antenna beamwidth, both horizontal and vertical
antenna beam pattern (beyond the mainlobe)
range
azimuth
elevation angle
Doppler velocity
time

Some of the parameters in the above list are easily measured while others require considerable time and effort. Special instruments are available for some of these while others depend upon the radar processor for providing correct values. As a meteorologist, you may not be able to make all of the required measurements and will have to depend upon the kindness of strangers to help you (i.e., engineers and/or technicians). I have heard it said that a good scientist doesn't trust anyone -- even themselves. Even if you must depend on the measurements of others, you should try to verify them. Further, there are often several ways to measure the same parameter. You should try to use more than one method just to insure the accuracy and precision of the values you use. Now, let us consider each of the above parameters individually.

And sometimes we overlook the obvious. I've added "time" to the bottom of the list. Time is easily measured nowadays, but it is not always correct! There have been numerous problems with radars or other data sets giving the wrong time. You should make sure that the time on your radar is correct. Compared to what? There are several ways to check time. It used to be a simple local phone call to the phone company to get time. That service may still be available in some locations, but are you sure *they* have the right time? Why not go straight to the horses mouth? In the United States, correct time can be obtained by listening to radio station WWV, the National Institute of Standards and Technology. Time is broadcast twenty-four hours a day at 2.5, 5, 10, 15 and 20 MHz. In addition to time (to the nearest thousandth of a second or so), WWV also broadcasts other information such as solar activity. The same information can be obtained by telephone by calling 303-499-7111. Another good source of accurate time is now available over the Internet. The U. S. Naval Observatory provides good time measurements from their homepage at "http://tycho.usno.navy.mil/cgi-bin/anim". This gives 30 seconds of continuously updating time. Normally it is good enough to get time to within a second or so, but occasional delays on the Internet can delay it or cause it to skip seconds. Even so, it is a convenient way to get time while sitting at your computer.

Directional couplers

In the radar block diagrams shown in the preceding chapters, the only components included were those absolutely necessary to make a simple radar understandable. Each of the items included (transmitter, waveguide, t-r switch, receiver, etc.) is itself composed of individual components. To understand how to make some of the measurements necessary to turn raw radar data into quantitative, meteorologically useful numbers requires that we add a little to the glittering generalities already given.

The waveguide shown earlier connects the various radio frequency (RF) components of the system. It did not allow for any external connections. Almost all radars, however, have some built-in devices which allow us to measure various components of the radar system by connecting into the waveguide system. One such connector added to the waveguide is called a directional coupler. Figure 12.1 shows the location of a directional coupler in a radar system. Basically, a directional coupler is a device which allows a small amount of energy from the main waveguide to be directed off into the coupler. Radars frequently have more than one directional coupler.

The amount of power that goes from one side of the directional coupler to the other (e.g., from the waveguide to the port or from the port to the waveguide, depending upon how it is being used) is usually on the order of 1/1000th or less of the total power. On a logarithmic scale, directional couplers typically have loss factors on the order of 35 dB or so.

Transmitter power

The power transmitted by a radar is measured through the use of a directional coupler and a power meter connected as shown in Fig. 12.1. The power meter reads the power at the coupler and gives the reading in mW or dBm. By knowing the losses of the directional coupler and the cable connecting the power meter and the coupler, we can then determine the power flowing past the directional coupler within the waveguide. We are really interested in knowing the power transmitted at the antenna (or just outside the radome, if the radar has one). To get this value, we also need to account for the losses in the waveguide between the directional coupler and the antenna (or through the radome).

Some power meters measure power by placing a thermistor (called a bolometer) at the end of the cable right near the directional coupler or pad. The amount of warming of the thermistor is related to the power being transmitted. In pulse radars, the transmitter is only on for a short time (typically about a microsecond) and then off for a moderately long time (typically a millisecond). The thermistor of the power meter cannot be heated and cooled that rapidly, however. Instead, it heats up in response to the *average* power applied to it. The average power is the power the radar would have to transmit continuously to produce the same average power that the real radar achieves by transmitting a high amount of power for a short time and then no power at all for a long time. The average power is related to the peak transmitted power through the following equation:

$$p_{ave} = p_t f \qquad (12.1)$$

where p_t is the peak power transmitted, p_{ave} is the average power, and f is the duty cycle of the radar. The duty cycle is the fraction of all time that the transmitter is actually transmitting. It is given by

$$f = \tau PRF \qquad (12.2)$$

where τ is the transmitter's pulse duration, and PRF is the pulse repetition frequency of transmitted pulses. For the typical values cited above, i.e., $\tau = 1$ μs and PRF = 1000 Hz, the duty cycle = 10^{-3}. Note that f is a unitless quantity. We frequently convert duty cycle into a logarithmic value; the example given then becomes a duty cycle of -30 dB. Consequently, for our example, the average power would be 1/1000 or 30 dB less than

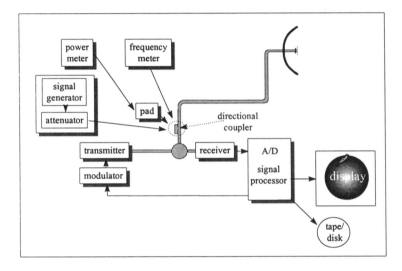

Figure 12.1 Block diagram of a radar showing the location of the directional coupler and where a signal generator and/or a power meter would be connected to make measurements of various system parameters. "Pad" is a fixed attenuator (not always needed) inserted between the power meter and the directional coupler to protect the power meter. The step attenuator is usually built into the signal generator. The analog-to-digital converter (A/D) is an integral part of the data processor.

the peak power. So, knowing the pulse duration, the PRF, and the average power (measured by the power meter), we can calculate the peak power transmitted.

The pad shown on Fig. 12.1 is there to protect the sensitive power meter. It is not always needed. If used, its attenuation must be accounted for in converting measured power into peak power transmitted.

We also have to correct for directional coupler, cable, waveguide, and radome losses to get the power transmitted into space from the radar. This is easily done in working with logarithmic measurements because losses are added together. Thus,

$$P_t = P_t' + L_d + L_c - L_P - L_w - L_r \tag{12.2}$$

where L_d, L_c, L_P, L_w and L_r are the losses (in dB) for the directional coupler, cable, pad, waveguide and radome, respectively, and P_t' is the peak power transmitted as measured at the directional coupler (given in dBm). Note in this equation that the first two loss terms are added and the last two are subtracted. Basically, we are trying to find the power inside the waveguide near the directional coupler. The power measured has been reduced by both the cable and directional coupler losses. Once it leaves there, however, it is then reduced by the waveguide and radome losses.

Transmitted powers vary from one radar to another, but are usually in the range from a few kilowatts to a megawatt or two. On a logarithmic scale, they range from perhaps 60 dBm to as much as 90 dBm or more. Average powers are typically about 30 dB less than these, depending, of course, on the PRF and pulse lengths used. The uncertainty associated with transmitted powers is usually on the order of 0.5 to 1 dB, depending upon the quality of the power meter and the care with which the measurements are made.

Transmitter frequency

The same directional coupler used to measure the transmitter power can be used to measure the transmitter frequency. In this case, we simply connect a frequency meter to this port and measure the frequency. Modern frequency meters are quite simple and accurate to use. In fact, of all the parameters in the radar equation, we probably know the transmitted frequency more precisely than any other. Power meters usually give frequencies to the nearest MHz or better. For a C-band radar, for example, with a frequency of 5500 MHz, an uncertainty of 1 MHz corresponds to an uncertainty of about 0.02%. On a logarithmic basis, this would be an uncertainty of only 0.08 dB. This is excellent compared to some of the other uncertainties in radar measurements.

Transmitter frequency is converted into wavelength λ through

$$\lambda = \frac{c}{f} \qquad (12.4)$$

where f is the transmitted frequency and c is the speed of light. We know the speed of light even more precisely than we know the frequency (see Chapter 3), so we also know wavelength quite well. The 0.02% uncertainty in frequency would apply to wavelength.

Received power

Most parameters used in the radar equation have a single value only. Thus, it is correct to say that we "measure" the parameter. Received power, on the other hand, can have a wide range of possible values. Further, it is usually not possible to make measurements of power from meteorological targets as they are being made because the radar scans past an individual target very quickly, both in range (i.e., at the speed of light!) and in azimuth.

How, then, can we measure the power received by a radar? Since the signals received from real targets are so fleeting, one way to do it would be to note how strong a target is on an A-scope, for example, and then later substitute an artificial signal of known, measurable power and adjust this signal to match that of the real target. Conceptually, this can be done fairly easily. When scanning thousands of points in space nearly simultaneously, however, this becomes an impractical technique for real-time data collection.

The device used for this substitution measurement is called a signal generator. Signal generators are available for all frequencies used by radars. The signals from these are injected into the waveguide using our friendly directional coupler. The signal from the generator goes into the waveguide and makes its way into the radar receiver where it is detected. This detected signal can be displayed on a PPI, RHI, A-scope or any other display used with the radar, or it can be processed and recorded along with the signals from real targets. On an A-scope we can watch the signal and adjust the power coming out of the generator to exactly match the strength of a real target.

The scheme outlined above is certainly possible, but it, too, would become quite cumbersome to apply to a large number of targets. By modifying this scheme somewhat and doing it ahead of time (or after the fact), we can generate enough receiver powers to cover the range of all possible powers we might expect the radar to detect. This procedure is not just a simple measurement anymore but is really a "calibration" of the radar receiver. In fact, the receiver is usually the only component of a radar that is "calibrated." Calibration is defined as the process whereby a position on the scale of an instrument is identified with the magnitude of the signal actuating that instrument (*Glossary of Meteorology*). It applies to the conversion of an input signal into a different output signal. As such, a calibration should span the whole range of possible input and output values using what is sometimes called a transfer function (i.e., the relationship between input and output signals). The term calibration is sometimes applied erroneously to antenna gain measurements and perhaps to other measurements, but it should be used correctly.

To perform a receiver calibration, a signal generator is connected to the directional coupler and turned on and allowed to warm up. The radar antenna should be aimed away from the ground and any other possible sources of echoes. It may be better to do the receiver calibration with the transmitter off. The frequency of the signal generator must be matched to that of the radar receiver. Further, you should make sure that the receiver is correctly tuned to the same frequency as the transmitter. There have been instances where receivers have not been tuned to the transmitted frequency for as long as two years, causing received powers to be off by 10 to 12 dB (Rinehart, 1978)!

The output of the signal generator should be stepped through known output powers that range from below the minimum- to above the maximum-detectable power of the receiver. You need to be careful at the high power end, however, so that you do not introduce so much power into the receiver that you cause damage. The signals injected into the receiver can then be recorded in exactly the same way all radar signals are recorded. In the early days of radar, scope photography was the preferred (sometimes the only) way to record data. Modern radars record data on magnetic tape or through other computer techniques. Other radars do not record any data at all but do display intensities on color displays. No matter how the data get used, the calibration procedure can be used to verify the quantitative aspects of the receiver.

For radars that use computers to do some of the data processing and recording, the power going into the receiver is usually converted into a digital signal somewhere in the system. This conversion takes place in an "A to D" converter (analog to digital converter). It takes the voltage (i.e., an analog signal) from the output of the receiver and converts it into a number or count or quanta (i.e., a digital signal). The range of possible output quanta depends upon the A/D converter used. A/D converters often produce 2^8 to 2^{10} possible values. For an "8-bit" A/D converter, the output quanta range from 0 to 255 (i.e., $2^8 - 1$). If such an A/D converter is used with a radar receiver that has a dynamic range of 90 dB, it would give a resolution of about 0.35 dB per quanta. Figure 12.2 shows an example of a radar receiver calibration for the NCAR CP2 S-band radar for 12 June 1981. The actual calibration values on the figure are the dots.

One term used above needs some amplification. That is the term "dynamic range." The dynamic range of a receiver is the difference (usually in decibels) between the minimum detectable signal or power and the saturation power. The minimum detectable power

(*MDS*) is the smallest power that can be detectable above the noise produced by the system itself. Some radars are incredibly sensitive. *MDS*'s on the order of -100 to -110 dBm are quite common on typical weather radars. Some even have *MDS*'s as low as -120 dBm or less.

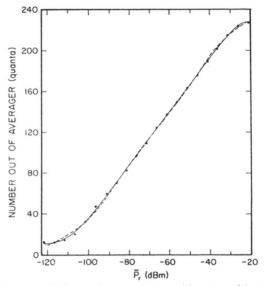

Figure 12.2 Radar receiver calibration (dots) for the NCAR CP2 S-band radar for 12 June 1981. P_r is the average received power going into the receiver (via the directional coupler). The dashed curve is the fit from a fifth-degree polynomial while the solid curve is the fit from the behavioral model.

Radar receivers are designed to operate over a range of powers. As the power going into a receiver increases, the output signal increases. However, if the input power is increased above some level, the receiver cannot put out any more power, and the receiver is said to be saturated. No matter how much the input signal is increased, the output remains the same. If too much power is put into the receiver, it can damage the receiver and burn out parts of it. Saturation powers (P_{sat}) for typical radar receivers are on the order of -30 dBm or so. Typical dynamic ranges of modern radar receivers are now on the order of 80 to 90 dB.

Now, returning to the problem of calibrating a receiver, one thing that must be done is to generate a set of output quanta as a function of input powers. One possible way of doing this is to change the input powers in such small steps (about 1/3 dB each) that every possible quanta from the A/D converter is covered between MDS and P_{sat}. This would require using an unreasonably large number of signal generator steps, i.e., 256 points for an 8-bit A/D converter. Besides, many signal generators have attenuators on them that only allow stepping the output power in 1 dB or 10 dB steps; steps smaller than 1 dB may not be possible.

An alternative technique is to step the signal generator through a range of values using steps on the order of 1 dB or perhaps 5 dB (Fig. 12.2 used 5-dB steps) and then fitting the resulting calibration data with some kind of equation or function to fill the gaps between the steps that were used. This equation can then be solved for input power, and a table of input powers versus all possible output quanta can be generated. This table can be stored as a look-up table that is quickly accessed to convert each measured output quanta into values of received power during data collection.

What kind of equation should be fit to the calibration data? Several possibilities exist. One commonly used class of equations is the polynomial of the form

$$P_r = \sum_{i=0}^{n}(a_i\, q^i) \qquad (12.5)$$

where q is the output quanta from the A/D converter, the a's are empirically-determined coefficients, n is the degree of the polynomial, and the summation, indicated by "$\Sigma(...)$", is carried out for n terms. To get a reasonably good fit to a set of calibration data, it is not unusual to have to use polynomial equations as high as the fifth degree. The resulting polynomial equation is then used to generate a look-up table of all possible quanta. The dashed curve on Fig. 12.2 is a fifth-degree polynomial fit to the calibration data. Note that the expression above for P_r is usually applied to logarithmic powers measured in dBm.

An alternative to a polynomial equation is to use a "piece-wise" linear fit. In this procedure the gaps between the calibration data points are fit with short straight lines, and the assumption is made that the receiver behaves reasonably between these points. This piece-wise fit is also used to generate a look-up table of all possible quanta. A disadvantage of the piece-wise fit is that if one of the data points is off for one reason or another, the surrounding points will also be slightly in error. Other techniques which smooth the data avoid this rather easily.

Another alternative to fitting calibration data is to use what has been called a "behavioral model" (Pike and Rinehart, 1983). This kind of equation is based on a consideration of the physics involved in a receiver. Briefly, it uses knowledge of how the receiver operates. At the small-signal end of the receiver's response range, there are two sources of input signals. One is the signal you are trying to measure and the other is that of noise detected and generated by the radar itself. The total power going into the receiver is the sum of these two powers. That is,

$$p_i = p_r + p_n \qquad (12.6)$$

where p_i is the total power into the receiver, p_r is the power received by the radar from a particular target, and p_n is the noise power from the receiver. All of these are linear powers measured in a consistent set of units such as mW or μW; mW are most frequently used. The noise power is similar to the minimum detectable signal of the receiver but is slightly different. It is the power generated by the receiver. It is often possible to detect signals that are actually *weaker* than p_n. This is possible by doing some averaging of the input signal; this tends to average out the noise and retain the signal. Noise, by definition, is random; consequently, averaging works quite well to minimize the effects of noise. True signals are not random; averaging does not make them disappear.

Many modern meteorological radars use what are known as logarithmic receivers. That is, the output of the receiver is proportional to the *logarithm* of the input power. If the receiver operates as correctly, the output is almost exactly logarithmic. When plotted on semilog paper (i.e., power on a logarithmic scale, quanta on a linear scale), the major part of the receiver response curve will be a straight line. Alternatively, if we plot the input power in logarithmic units on a linear scale, the receiver response curve will still be a straight line. Thus, using logarithmic input powers (i.e., powers measured in dBm), the middle of the calibration curve will be a straight line, and we can use linear regression to fit a linear equation to this part of the curve.

Chapter 12

The remaining important part of a receiver's calibration curve is at the saturation end. As we approach saturation from the middle of the calibration curve, an input signal (in dBm) will initially continue to cause a linear increase in the output quanta. As we get closer to saturation, however, an increase in input power starts to produce progressively smaller increases in output quanta. Finally, we reach the point where further increases in input power cannot change the output signal at all. This kind of response can be modeled by an expression of the form

$$\frac{p_s}{p_r + p_s} \tag{12.7}$$

where p_r is again received signal power and p_s is the saturation power. Both p_r and p_s are measured in linear power units such as mW.

If we combine all three of these effects (noise, logarithmic output of the receiver, and saturation), we can get the following "behavioral model" for a radar receiver:

$$10 \log_{10}\left((p_r + p_n)\left(\frac{p_s}{p_r + p_s} \right) \right) = a_0 + a_1 q \tag{12.8}$$

where the coefficients a_0 and a_1 are empirically-determined linear-regression coefficients obtained by regressing the entire left side against the output quanta q.

There are a couple of flies in the ointment above. In order to use the Eq. 12.8, we need specific values for p_n and p_s. They could be measured independently and inserted into the equation. This is perfectly acceptable.

However, they can also be obtained from the calibration data set itself. To do this for p_n, for example, we could use the bottom half of the calibration curve (the end affected by noise) and iteratively try many values of p_n. We then choose the one which has the best fit as determined by looking at the coefficient of determination between the calibration data and the predicted values obtained from the fit of the data. Similarly, using the top half of the data near saturation, we can iteratively determine p_s. The description of this may sound imposing, but this procedure can easily be automated such that it is simply a matter of entering the table of calibration data and running a simple program. We then get objectively-determined values for both p_n and p_s.

A third way to determine p_n and p_s is to draw a graph of the calibration data such as in Fig. 12.2 and fit a straight line to the middle portion of the calibration curve. p_n is the point where this straight line intersects the straight horizontal line drawn through the noise-level end of the curve. Similarly, p_s is the found at the intersection of the straight middle part of the calibration curve and the straight horizontal line drawn through the saturation-level data. Figures 12.3 and 12.4 show the determination of noise power P_n and saturation power P_s, respectively.

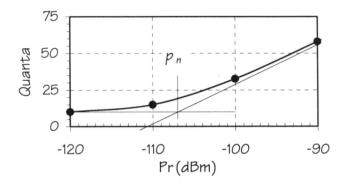

Figure 12.3 The noise-end of a receiver calibration curve. The dots and the smoothed curve are the calibration data. The diagonal line is a straight-line fit to the logarithmic portion of the calibration curve. The horizontal line is a fit to the noise-power end of the calibration curve. The noise power p_n is found where the straight lines intersect.

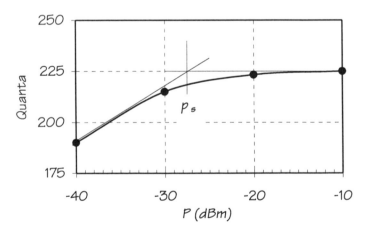

Figure 12.4 Similar to Fig. 12.3 but for the saturation end of a receiver calibration curve. P_s is the saturation power of the receiver and is found where the horizontal and diagonal straight lines intersect.

Once the values for p_n and p_s are available, the whole term on the left is only a function of p_r. Thus, it is a simple matter of linear regression to get a_0 and a_1 as a function of q and p_r. Then, given the complete calibration equation, we can generate a look-up table of powers for all possible quanta values. The solid curve on Fig. 12.2 is the result of the behavioral model fit to the radar receiver calibration data.

Occasionally a radar logarithmic receiver will be found that does not have a perfectly straight output signal as a function of input power. If this happens and if even better calibration results are needed, it would be possible to combine the behavior model and a polynomial expression to accommodate nearly any calibration curve shape. The right side of the calibration equation could be a polynomial while the left side would remain unchanged. The left side would still accommodate the physics of the noise and saturation ends of the calibration curve while the right side would accommodate the individual personality of the middle part.

No matter what method is used to generate the analytical expression or the look-up table for the receiver calibration, the use of the calibration is the same. As the radar receiver and its processor operates on real data, the A/D converter will continue to provide quanta at each measurement point (range gate and azimuth). The power corresponding to the output quanta are determined by entering a look-up table. These powers are then entered into a simple form of the radar equation such as was given in Chapter 4. In this case it is solved for radar reflectivity factor, for example, and is given as

$$Z_e = P_r + 20 \log r + C \qquad (12.9)$$

249

where P_r is obtained from the receiver calibration look-up table (in dBm), r is range to the particular range gate (usually in km), and C is a constant for the radar (in decibels). (See Appendix D for radar constants for several different radars.) It may also be desirable to use a look-up table for "20 log r" since calculating logarithms with computers is moderately time consuming; besides, the ranges used are always the same for a given mode of operation. Once calculated, the reflectivities can then be recorded on tape and/or displayed as colors on the displays.

As a final note on the use of the behavioral model to fit a radar receiver calibration, we can compare the quality of the fit of the behavioral model and the fifth-degree polynomial to the calibration data shown in Fig. 12.2. One measure of the quality of a fit is to calculate the average difference between the actual calibration data and the predicted powers for a given set of quanta (Pike and Rinehart, 1983). This average difference is called the residuals of the fit and can be measured in decibels. For the data in Fig. 12.2, the variance of the residuals for the fifth-degree polynomial was 3.7 dB² while the variance of the residuals of the behavioral model was 2.3 dB². This indicates that the behavioral model fit the data a fair amount better than did the arbitrary polynomial. In fact, the behavioral model fit this particular calibration better than polynomials up to a tenth degree.

Receiver bandwidth

Measuring the receiver bandwidth is not for amateurs. This is one place where you may have to trust the measurements your engineer or technician give you. Alternatively, you may have to use the values quoted by the manufacturer.

Note, however, that there is more than one bandwidth measurement you may need. Doviak and Zrnic' (1993) talk about both a noise bandwidth ("not so simply specified," they say) and a receiver filter bandwidth B_6 ("which is best specified as the frequencies within which the attenuation of power is less than one-fourth of its highest level"). Make sure you know which is which and when to use each.

The major use I have made of receiver bandwidth is to make antenna gain measurements using the sun. The sun radiates energy over a very wide range of frequencies. A radar receiver is sensitive to only some of these. The bandwidth of the receiver limits the frequencies to only a specific band. The amount of power in this band is what is detected and measured by the radar; all other frequencies are rejected and lost.

Pulse duration

The pulse duration or pulse length is another relatively simple parameter to measure. This can be measured directly on an oscilloscope by sampling the transmitted pulse via the directional coupler. There are other ways of measuring this, but the measurement on an oscilloscope is the easiest.

Another way to get an estimate of the transmitted pulse length is to measure the length of the echo from a point ground target such as a radio tower. An isolated radio tower (isolated so that no other ground targets are nearby to contaminate the measurement) should be much smaller than the pulse length. Consequently, the reflection of radar waves from it should exactly mirror the transmitted pulse. Unfortunately, this does not work perfectly well because the reflected signal must first go through the radar receiver before it can be measured. And radar receivers do not always respond perfectly well to rapidly changing input signals. Doviak and Zrnic' (1993) discuss this problem in some detail. The net result of the receiver response function is that the return from a point target is not quite the same shape and duration as the transmitted pulse. The general tendency is that the receiver will slightly lengthen the signal so that the "pulse length" measured from the echo of a point target will be somewhat longer than the length of the transmitted pulse. For example, the pulselength of the UND C-band radar is 0.6 μs while the length of a radio tower point target on an A-scope appears to be more like 1 μs long.

For pulse length, you may need to depend on the measurements made by the site engineer/technician. But you should also check the measurement yourself using a well selected ground target from time to time.

Antenna gain

The gain of an antenna is a property of the antenna itself. Its value depends upon the size and shape of the antenna and the wavelength of the radar. Once gain has been measured, it should remain constant as long as the antenna does not change. Because of this, it is worth investing the time to make a good antenna gain measurement at least once for each radar site. It is probably worth repeating this measurement occasionally to insure that everything is okay.

One reason this is advisable is that most antenna gain measurements are not just measurements of the gain of the antenna, but they also include waveguide losses between the antenna itself and the directional coupler used to measure transmitter power and receiver calibrations. If corrosion or something else takes place in this section of waveguide, it will change the effective gain of the system. By monitoring antenna gain occasionally, changes in its value should alert you to possible problems in the waveguide system.

There are several ways to obtain the gain of a radar antenna. The simplest is to use the value quoted by the manufacturer. If an antenna is new, this probably is acceptable. Manufacturers, however, sometimes quote the theoretical gain or the gain for which the antenna was designed rather than actually measuring the real gain of a specific antenna. This may mean that the value quoted does not really apply to your antenna. Since antenna gain usually appears as gain squared in the radar equation, an x-dB error in gain would produce a $2 \times$ -dB error in received power, reflectivity or backscattering cross-sectional area.

Antenna gain can be measured using signal generators (with standard horns), standard targets, the sun, or using an antenna range. This latter is quite expensive and requires moving the antenna to the antenna range, a time-consuming activity. Also, the act of removing an antenna and reinstalling it can change the shape of the antenna somewhat, so the gain of the antenna might be slightly different in operational use from what was measured at the antenna range.

For the signal generator/standard horn technique, the procedure is to place a signal generator at some distance from the radar, usually on a tower, building or hill, and aim it toward the radar. By knowing the power transmitted by the signal generator, the gain of its horn antenna, the distance between the radar and the signal source, and measuring the received power, we can calculate the gain of the antenna using a modified version of the point target radar equation. This technique usually takes two or more people an hour or more to perform. Communications between the radar and the signal generator site are also useful to coordinate the activities of the team performing the measurement. This same general procedure can be extended to measure the entire beam pattern of an antenna if this is needed.

Another way to measure antenna gain is to use a standard target. In this case the radar acts as the signal generator, sending out its normal transmitted power. The standard target is located at a known distance from the radar, and the power received from the target is measured. By knowing the backscattering cross-sectional area of the target, the transmitted power, the range to the target, and the received power, we can calculate antenna gain using the point target radar equation (Eq. 4.7). Similarly, by scanning this target in azimuth and in elevation, we can determine the antenna beam pattern, at least for sidelobes near the mainlobe.

The most commonly used standard target for antenna gain measurements is a metal sphere of 10- to 30-cm diameter. A metal sphere will give exactly the same return no matter how it is oriented relative to a radar. The sphere can either be tied to a balloon and released or it can be tied to a balloon and tethered in place. In the case of a released balloon/sphere, the target must be tracked and measured on the fly. This is not an easy task and tends to give poorer (and more expensive!) measurements than using a tethered balloon/sphere system.

In the case of the tethered balloon, the sphere is tied some distance below a helium (or hydrogen -- be careful!) filled balloon and these are raised to some distance above the surface. Typically, the balloon will have to rise to a distance of a hundred meters to perhaps several hundred meters. The location where the balloon is launched must be carefully chosen so that there is not a lot of other ground clutter in the area. At the same time, it must be in a location where the balloon/sphere can get high enough to be clearly detectable by the radar. A distance from the radar of from 1 to 10 km could be used. The closer to the radar the balloon is, the higher the elevation angle will be for the same height above the ground. However, if you are too close, you may not get a good measurement because radar receivers take a little time to recover from the massive transmitter pulse that usually leaks into the receiver every time the transmitter fires. This receiver "recovery" time can last for the equivalent distance of a couple hundred meters or more, depending upon the radar. You should be sure that the balloon site is far enough from the radar that receiver recovery time does not distort the measurements.

Antenna gain measurements using either free-floating or tethered spheres usually require a team of two people or more at the balloon site and one or more people at the radar. Good communications are essential between them. It usually takes an hour or more to perform a gain measurement using a sphere. One human problem related to gain measurements using spheres is that it is usually necessary to do these at what some people consider inhumane times of day, typically near sunrise, so that the winds will be light. When it is windy, it is difficult to impossible to safely launch a balloon, so this kind of measurement may have to be delayed until conditions are acceptable.

A second kind of standard target that has been used to make antenna gain measurements is a dihedral target (Rinehart and Eccles, 1976). In this case the target can be permanently mounted on top of a pole at some known location. The dihedral must be oriented so that its vertical fold line is exactly perpendicular to the radar mainlobe axis. Once this is done, it is then simply a matter of aiming the antenna at the target, measuring the power received from the target, and calculating antenna gain. For this gain measurement, you must know transmitted power, received power, distance to the target, and the backscattering cross-sectional area of the dihedral. The equation for this was given earlier.

There are two ways the dihedral has been used in the past. One was to have someone go out to the target and turn on a small motor located behind the dihedral target. This caused the reflector to nod back and forth slightly either side of being perpendicular to the radar beam. This gave a clear signature in the returned signal so that the people at the radar site could be positive that the correct target was being examined. When done this way, it would take one person at the target site and about half an hour or less to perform the gain measurement. The second way it was used was to leave the target set so it was perpendicular to the radar beam. Then it was simply a matter of aiming the radar at the target's fixed location and making a measurement of received power. When done this way, it took about 10 min to perform an antenna gain measurement.

Besides the advantage of being much less labor and time intensive to use the dihedral target to measure antenna gain, the dihedral has a couple of other advantages. It can be used in most weather conditions and any time of day or night. By having its location surveyed, we could use its position as a check on range, azimuth and, to a lesser extent, elevation. Thus, every time we use it, we could check several radar parameters.

Antenna gain using the sun

A final method that can be used to measure antenna gain is through measurements of the sun (Whiton et al.,

1976; Frush, 1984). The sun radiates not only visible light, it radiates electromagnetic energy at all frequencies. The amount of energy emitted by the sun at radar frequencies is sufficient to be detectable by most modern radar receivers. It is simply a matter of aiming the antenna at the sun and measuring the power received. Note that we do not need to use the transmitter for this. We are not bouncing an echo off the sun; we are using the sun as a "calibrated" signal generator at a known position.

There are at least two problems in using the sun to measure antenna gain. One is to know where the sun is, and the other is to know how much power the sun is radiating at any given time. The second problem is relatively easily solved. There are a number of solar observatories located around the world which measure the solar flux density at a number of different frequencies each day. Solar flux measurements are available by calling the Space Environmental Forecast Center, U.S. Department of Commerce, Boulder, Colorado, a day or more after the measurements are made (303-497-3171). You will have to specify the date and the frequencies you want. You will need data at frequencies both above and below your radar frequency (unless, of course, your radar happens to operate at exactly the same frequency as one of the measurements; this doesn't happen). Both Whiton et al. and Frush describe how to use solar flux measurements to get a value for use with your radar.

There may be an alternative to calling the Space Environmental Forecast Center to get the solar flux, and that is to get solar flux from the Internet. The whole reason for having to call to get the solar flux density in the first place is that the output of the sun is not constant but varies from time to time. But when the sun is "active", it is usually active proportionally at all frequencies, not just at a single frequency. So, if we can get the flux density at one frequency, we should be able to scale the flux at another frequency up or down proportionally. This involves two steps, getting a representative flux and knowing the relationship between that flux and the flux at our radar's frequency. The first step is easily solved by getting the solar flux density at 2800 MHz, a frequency which is reported daily over the Internet from the Solar Radio Programme, National Research Council, Dominion Radio Astrophysical Observatory, Penticton, British Columbia, Canada V2A 6K3:

http://www.drao.nrc.ca/icarus/www/current/current.flx

Using the 10.7-cm wavelength flux data at other frequencies is somewhat more of a problem. To do this easily, you would first have to find the flux at your frequency at one time along with the flux at 10.7 cm. Then you should be able to get the 10.7-cm flux and simply adjust to your frequency. This should give a value at your frequency that is within a decibel or so, and that is close enough much of the time. If nothing else, you could monitor the flux and, if it hasn't changed much, continue to use your existing value at your frequency. If the flux at 10.7-cm wavelength changes dramatically, then you might check for current values at your frequency and use the better value.[11]

---

[11] One additional use of the 10.7-cm flux value that is interesting is to use it as a proxy for sun spot number. The sun spot number is given as $N = 1.14 \cdot S - 73.21$ where S is the solar flux density in solar flux units (1 solar flux unit = 10^{-22} W m^{-2} Hz^{-1}).

The other problem, that of knowing the location of the sun, can be solved a number of ways or it can be looked up in a solar ephemeris or nautical almanac. Alternatively, you can go outside and see where the sun is and aim the antenna accordingly. If you systematically scan in the general vicinity of the sun, you should be able to detect the sun's signal on the radar. Once it has been approximately located, you can scan it in azimuth and elevation to get the sun exactly on the antenna beam axis where its signal will be strongest.

A simple way to locate the sun is to use what I like to call the "rule of fist"! If you hold your fist out at arm's length, the approximate angular size of your fist is 10°. By holding it vertically and sighting along the bottom of your fist to the horizon (or horizontally, if there are things above you on the horizon) and then counting fists up to the sun, you should be able to get the sun's elevation fairly accurately. And if the fist is 10°, each finger should be 2°. You can do the same thing in azimuth with your fist held horizontal. Of course, you will have to know some reference direction or point of the compass. You can always check the "calibration" of your fist by measuring from the horizon to the zenith: it should be nine fists!

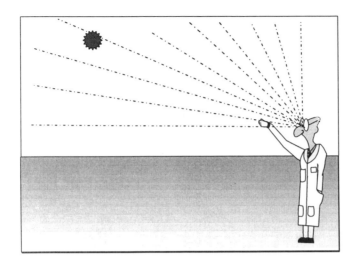

Figure 12.5 Using the rule of fist to measure the elevation angle to the sun.

A better way to get the sun's position is to calculate it. By doing it this way, you can check the azimuth and elevation angle of your antenna at the same time. In fact, the sun is a good target for aligning the azimuth of a radar. Its position is accurately predictable and provides a good target which can be used frequently for checking antenna alignment.

There are several equations needed to calculate the position of the sun. Without going into all of the background needed to explain the sun's position, we can state that the position of the sun depends upon time of day, day of the year, latitude and longitude (we can safely ignore height above sea level for all but the most exacting requirements).

The following set of equations are useful for getting the azimuth and elevation angle to the sun from any location at any time. They are not, however, as accurate as using almanac values. An excellent source for exact values is *The American Ephemeris and Nautical Almanac* which is published annually by the U.S. Government Printing Office, Washington, D.C. Another excellent source is *for DOS, 1990-1999, Version 1.0* (Multiyear Interactive Computer Almanac) published by Astronomical Applications Department, U. S. Naval Observatory 3450 Massachusetts Ave., N.W., Washington, D.C. 20392-5420). This program calculates the azimuth and elevation to the sun (as well as the moon, all the planets and lots of stars). It is an excellent way to get precise positions of heavenly bodies.

Another excellent source of equations to calculate the sun's azimuth and elevation angles is "The NOAA Solar Ephemeris Program" (Taylor, 1981). This program also contains complete FORTRAN listings of the required equations along with explanations of their use and the accuracies of the results. This program is the standard against which the following simplified equations are compared later.

The azimuth and elevation angles calculated by the equations to be given shortly are usually within about 0.5° of those calculated by MICA most of the time, sometimes much better than this. If you need better accuracy, you might consider using one of the above sources for your measurements.

The declination of the sun is the angle the sun is above or below the equator. Declination is given approximately by

$$\delta = -23.5° \cos\left((D + 10.3 \, d) \, \frac{360°}{365.25 \, d}\right) \quad (12.10)$$

where D is the ordinal day or day of the year, and δ is the declination angle (in degrees) above (+) or below (-) the equator. Notice that this equation does not account for leap year; it treats every year the same. This is one reason the azimuths and elevations calculated using this and following equations cannot compete with the almanac values for great precision.

The term on the right side of the above equation (360°/365.25 d) converts the angular term into degrees from days. If you are using a calculator or computer which is capable of using angles in units of degrees, this is all that is needed. If your computer requires that all angles be given in radians instead, you must substitute a term "2π rad/365.25 d" for the term "$360°/365.25$ d". The following equations assume you can use degrees or that you will convert to radians as needed.

The earth does not go around the sun in a circular orbit but in an elliptical orbit. One of the consequences of this is that its speed of travel around the sun changes, depending upon where earth is in its orbit. A consequence of this is that solar time changes slightly from day to day compared to clock time. To account for this, we need to correct our clock times by a term called the "equation of time"[12] so that we can get correct solar time. The equation of time is given approximately by

$$EQT = 0.123 \cos (time + 87°) - \frac{1}{6} \sin \left(\frac{time + 10°}{0.5} \right)$$

(12.11)

where EQT is in hours and "$time$" is the day of the year expressed in degrees, i.e.,

---

[12] "Equation of time" did not originally mean an equation at all. Rather, it was a graph of how much solar time (also called apparent time) differed from mean time (or standard or clock time).

$$time = D \frac{360°}{365.25\,d} \qquad (12.12)$$

EQT is given in hours. Again, this equation ignores leap-year variations. The equation of time and clock time will be combined shortly to give the local hour angle (*LHA*) in degrees for further calculations.

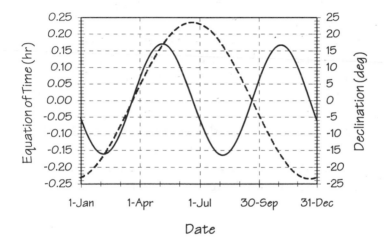

Figure 12.6 Equation of time (left axis and solid curve) and declination (right axis and dotted curve) as a function of day of the year.

Standard time zones are every 15° of longitude apart, starting at 0° longitude, the longitude of Greenwich, England. By convention, longitudes east of Greenwich are positive while those to the west are negative. North Americans frequently ignore this convention when expressing longitudes over the western hemisphere and simply append a "W" to the end of the longitude to indicate that it is west of 0°.

Thus, we would write the longitude of Grand Forks, North Dakota, as 97° 00' 38" W instead of -97° 0' 38". We do a similar thing with northern and southern latitudes. The center of the Eastern Standard Time zone (EST) is at 75° W; Central Time Zone (CST) is centered at 90° W longitude; Mountain Standard Time (MST), at 105°W, and Pacific Standard Time (PST), at 120° W. Be sure you use western longitudes as negative angles in MICA or any other program that calculates solar positions.

Solar time differs from clock time by an amount that depends upon our longitude. If we are east of the standard longitude, the sun will be directly south of us (local noon) before it is local noon at the standard longitude. Since each time zone is 1 h apart (15° of longitude apart), each degree of longitude equals 4 min of time. If we are west of the standard longitude, local noon will be later than on the standard longitude. To correct for this we need to get the angular distance between the standard longitude for the time zone we are in and our own longitude and convert this into time:

$$\delta t_\lambda = (\lambda_{std} - \lambda)\, \frac{1\,h}{15°} \tag{12.13}$$

where λ_{std} is the standard longitude (e.g., 75°, 90°, etc.), λ is the longitude of the radar, and δt_λ is the time correction (in hours) for our longitude.

Now we need to combine the equation of time, the longitudinal correction factor and convert to what is called the local hour angle. This is done using

$$LHA = (local\ clock\ time + \delta t_\lambda + EQT + 12\,h)\, \frac{360°}{24\,h} \tag{12.14}$$

where all of the terms in the parentheses on the right are expressed in hours. LHA is expressed in degrees. If LHA is greater than 360° or less than 0° (i.e., greater than 24 h or less than 0 h), we subtract or add 360° to it to keep it between 0° and 360°.

The azimuth Az and elevation angle el of the sun are calculated using

$$sin\ el = cos\ z = sin\ \phi\ sin\ \delta + cos\ \phi\ cos\ \delta\ cos\ LHA \quad (12.15)$$

$$x = tan\ Az = \frac{sin\ LHA}{cos\ LHA\ sin\ \phi - tan\ \delta\ cos\ \phi} \quad (12.16)$$

where ϕ is the latitude of the radar (positive if north of the equator, negative if south), δ is the declination of the sun, and z is the angle from the zenith to the sun (i.e., from straight overhead down to the sun; elevation angle, on the other hand, is the angle from the horizon up to the sun). Incidentally, these are exactly the same equations suggested in another U.S. Naval Observatory reference (*Almanac for Computers*) for calculating the azimuth and elevation of celestial objects. Thus, any difference between our values and their values is a result of their use of more exact approximations for declination and the equation of time.

Since computers and calculators normally give the arctangent in the range -90° to +90°, the correct quadrant for Az can be selected according to the following rules:

If 0° ≤ LHA < 180°,

Az = 180° + arctan x, if x is positive

Az = 360° + arctan x, if x is negative.

If $180° \leq LHA < 360°$,

Az = arctan x, if x is positive

Az = $180°$ + arctan x, if x is negative.

In the above equations Az is the azimuth angle to the sun measured clockwise from north (north = $0°$, east = $90°$, etc.) If sin el is negative, the sun is below the horizon. Note that these equations do not account for the refraction that normally takes place in the atmosphere; when the sun is low in the sky (below about 10 to 15°), the angles calculated, particularly elevation angle, may be slightly in error.

Equation 12.15 has one additional use that is fun to utilize. If we set the elevation angle to zero and solve for the time, we end up with an equation that can be used to predict sunrise and sunset times. Doing this give

$$SRtime = \cos^{-1}\left(-\tan\phi\tan\delta\right)\frac{1h}{15°} - \delta t_\lambda - EQT - 12h$$
$$(12.17)$$

$$SS\ time = 24\ h - SR\ time \qquad (12.18)$$

where, again, all of the terms on the right sides of Eqs. 12.17 and 12.18 are in hours.

One sometimes vexing complication in all of the above calculations is that of daylight savings time. The equations given above are correct for standard times. During the approximately six months when we operate on daylight savings time, our watches read one hour faster than standard time. Consequently, we must either change the equations above to subtract an hour from our clock time or we must make this subtraction before we enter time into them. In either case, be sure to do it correctly. One simple check to insure you did it correctly is to calculate the azimuth of the sun for a time of 12:00:00. At that time the sun should be close to an azimuth of 180°, depending, of course, upon your actual longitude and the day of the year (because of the slight changes the equation of time introduces into this calculation). If the calculated azimuth differs too much from 180°, try entering a time of 1100 or 1300 and see what happens.

All of the above equations seem rather formidable at first sight. They should not be. It is possible to squeeze all of these equations into a relatively small program. I have these in a programmable pocket calculator. By entering the latitude, longitude, day of the year, and time, it quickly gives the equation of time (minutes), declination of the sun, sunrise and sunset times, and the azimuth and elevation angles to the sun. This is quite handy for pointing radars at the sun to make quick system checks. [Actually, if the truth were known, I do this as much for the fun of detecting the sun with a radar as for any deep scientific purpose.]

Okay, now that we can determine the position of the sun, how do we use it to get antenna gain? The primary references for this are the Frush (1984) and Whiton et al. (1976) cited earlier. Basically, the method consists of measuring the power detected by the sun and correcting it for several factors. These factors include those for: polarization, earth-sun distance, antenna size, receiver bandwidth, and the extended-target effect.

A radar detects only one polarization component whereas the sun radiates all polarizations. So, we have to increase our measured power by a factor of two.

A second factor to consider is that the earth's orbit is an ellipse, not a circle. At some times during the year the earth is closer to the sun than at other times. When we get published solar flux values from observatories, they are usually corrected for a distance of 1 astronomical unit (the average distance from earth to sun). To compare our measurements to observatory measurements, we also need to make this correction. Castilli, *et al.* (1975) give this correction in tabular form.

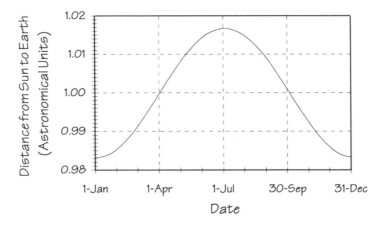

Figure 12.7 Distance from earth to the sun as a function of day of the year (measured in astronomical units)

Thirdly, the sun is neither a point target nor a beam-filling target. To correct for this, we need to use a correction factor given by Frush.

Fourthly, observatory measurements are corrected for atmospheric attenuation so they will be the equivalent to what would have been measured above the atmosphere. We need to correct our measurements for this same effect.

Once all of these corrections have been made, we should be able to use the solar flux density on a given day to calculate the gain of our radar's antenna. Once most of the corrections have been calculated once, they will not change, so it becomes relatively easy to make measurements of the power received from the sun and check overall system and antenna gain quite easily.

Antenna beam pattern

There are also a number of ways the beamwidth of an antenna can be measured. As already mentioned, these methods are essentially the same as for measuring gain except that the antenna scans the signal source in azimuth, elevation, or both. By recording data during the scanning procedure, a complete beam pattern can be produced. Of course, the best pattern would be obtained by moving the antenna to an antenna range, but this is usually not worth the effort. So, the signal generator or standard target may be used instead. The balloon-borne sphere probably would not be very good for this for at least two reasons. One is that the return from the sphere may not be strong enough to cover the entire sidelobe pattern. And second, wind might cause the balloon to drift in position enough that errors would be introduced in the measurements.

Since the antenna beam pattern is a relative measurement (one that does not require an absolute antenna gain value), it is not necessary that we know the actual backscattering cross-sectional area of the target being used. All that is necessary is that the target give a strong but steady return. Many radio towers or other point targets may meet these requirements. Rinehart and Frush (1983) described a technique for using a ground target to get the complete antenna beam pattern for a radar.

The procedure to do this is as follows. Select a strong, isolated ground target as an echo source. Scan the target slowly in azimuth from as low as the antenna will go to perhaps 10° above the target or higher, recording data continuously. By scanning slowly enough, you should be able to have an azimuthal resolution on the order of 0.1° or so. At the end of each azimuthal scan, raise the elevation of the antenna by 0.1° to 0.2° and repeat the azimuthal scan. Continue doing this until the highest desired elevation angle has been reached. The strongest reflectivity or power received will be the "on-axis" signal; all others can be compared to this value to get the "relative antenna gain" at all azimuths and elevations.

A reasonable estimate of the antenna beam pattern will be available for the single range gate containing the ground target with little or no further processing. However, since the aforementioned scanning may take as long as 30 min or so, it is possible that slight changes in transmitted power may introduce errors into the data set. This may need to be accounted for before using the pattern obtained. Also, birds and insects flying near the target (at the same range) may contaminate the data occasionally. These can be edited out manually, since they will usually be evident because they will not exist at adjacent elevation angles; true sidelobe features are usually much more ·than 0.1° across. If desired, the data collection procedure can be repeated more than once and the results combined together. When done, you should have the data to generate a beam pattern that will resemble Fig. 2.9.

Antenna beamwidth

The antenna beamwidth is the angular width of the mainlobe of the antenna beam pattern at the point where the gain is exactly one half of the on-axis gain (i.e., 3 dB below the maximum). However, if the gain measurements are made by transmitting a signal from the radar to a fixed target and receiving the echo back, then gain is in the radar equation twice (i.e., g^2). In this case we have measurements which include the "two-way" pattern, not just the "one-way" pattern. Here we have to go 6 dB down from the peak to find the beamwidth. In both cases, the beamwidth should be the same.

The beam pattern obtained from a ground target is perfectly suited for getting the beamwidth. Simply choose the elevation angle which gives the strongest return; that, presumably, is directly through the center of the beam axis. Usually this will be the lowest elevation angle, but in some cases your radar may be able to scan below the strongest part of the target. Then you would need to use whichever elevation angle gives the strongest return.

The vertical beamwidth can be obtained by scanning upward through a strong ground target. In this case you may obtain only half of the beam pattern. That is, if the target is at the same elevation as the radar antenna, then 0° elevation angle should give the strongest return. It may not be possible to scan below 0° with some radars. In this case, simply find the angular distance between 0° and where the gain is 6 dB lower than the maximum (for two-way gain measurements), and double it. If the antenna cannot scan in the vertical direction (NEXRAD radars have only azimuthal scan capability), elevation data can be composited by always using the azimuth which has the strongest return at the *bottom* elevation angle. If necessary, interpolate to find this value. While not as satisfying as a direct vertical scan through a target, it would give a reasonable check on the vertical antenna beamwidth value.

Range

There is usually not a problem in making range measurements with radars. Radars are, after all, named for this capability (radio detection and RANGING). Nevertheless, there can be problems in making range measurements.

Modern radars which use range gates for collecting data at specific distances from the radar have distance calculated based upon an internal clock of some kind. The master clock which tells the transmitter when to transmit probably also tells the range gate counter when to start counting. On some radars, it takes a short period of time to generate the transmitted signal. Depending upon how long this delay is, there can be an error in range introduced into the measurements. With some radars, for example, as many as two or three range gates can tick off before the radar ever transmits! Thus, it is highly desirable to check range occasionally.

There are a couple of ways to do this. One easy way is to use ground targets of known location. The distance to these can be measured on an A-scope, or, preferably, they can be verified in the recorded data that has passed completely through the radar data processor. Several targets should be used, not just one. Averaging the differences between known and measured distances (and azimuths) for several targets should give better answers for range (and azimuth) errors than would a single measurement.

But how do we know the "real" position of a ground target? One way is to have their locations determined by a survey. $$$! Another way is to use U.S. Geological Survey topographic maps. Maps of 1:24 000 or 1:25 000 should be available for almost every location in the United States. These maps show the locations of radio towers, water towers, power plants with large smoke stacks, and a lot of other potentially useful targets. Unfortunately, some of these maps are outdated, so beware that all the towers shown on the maps may no longer be there or towers that do exist now may not be on the maps.

Given a good map and a good target, how can we get the range and azimuth from the radar to the target? Obviously, we can measure these with the radar, but it is exactly these measurements we are trying to check. One way to do this is to simply measure the distance and azimuth on the map itself. This works okay if the radar and the target are on the same map. Frequently, however, the best targets are on another map. Adjacent maps can be taped together to form larger maps, and the measurements can proceed as before. If the target is too far away, however, taping maps together can become unwieldy and less accurate.

An alternative to this is to use what is called great-circle navigation equations to calculate the distance and heading between any two points with known latitudes and longitudes. To use these, it is necessary to determine the latitude and longitude of the radar and of each target from topographic maps. Then the heading and range are calculated using the following equations:

$$R = cos^{-1} \left(sin\ \phi_r\ sin\ \phi_t + cos\ \phi_r\ cos\ \phi_t\ cos\ (\lambda_t - \lambda_r) \right) F$$
(12.17)

$$\Theta = \cos^{-1}\left(\frac{\sin \phi_t - \cos (R / F) \sin \phi_r}{\sin (R / F) \cos \phi_r}\right) \quad (12.18)$$

where ϕ is latitude, λ is longitude, subscripts r and t are for the "radar" and the "target", respectively, F is a unit conversion factor, R is distance, and Θ is heading. If $\sin(\lambda_r - \lambda_t) < 0$, then $\Theta = 360° - \Theta$.

The unit conversion factor F is the distance along the earth's surface represented by a change in latitude of $1°$. It is given by

$$F = \frac{C}{360°}$$

where C is the circumference of the earth

$$C = 2 \pi R_e$$

and R_e is the radius of the earth. Unfortunately, the earth is not a perfect sphere but is slightly fatter around the equator than around the poles (not unlike middle-aged people). The equatorial and polar radii of the earth are 6378.39 km and 6356.91 km, respectively. At a latitude of 40°, the equivalent earth's radius is about 6374.2 km. This corresponds to a value of $F = 111.250$ km/° latitude. For most practical purposes in the continental United States, this value should be perfectly fine. But if your radar is near the equator or one of the poles, you might choose a better value to use.

One slight problem relates to the speed of light. As mentioned more than once, this is a very well known constant. For many practical purposes we can approximate its value from 299 792 458 m/s to 300 000 000 m/s. If extremely accurate ranges are need at long distances from a radar (as, for example, in trying to locate a specific range gate relative to a specific raingage or aircraft position), it might be necessary to use the best estimate of the speed of light to calculate this distance rather than the approximate value. At a range of 150 km, for example, the rounded value would be in error by about 100 m. Another few tens of meters of error result from ignoring the fact that the speed of light in the atmosphere is slower than it is in outer space by about 0.03%. Almost everyone ignores these errors -- and justifiably so (most of the time).

A good way to get the latitude and longitude of a ground target is to use a portable GPS (global positioning system) device. These can be carried from target to target to provide ground positions that should be accurate to within a few tens of meters or so. Since most radar pulse lengths are on the order of 100 m or longer, this is fine, especially if several targets are averaged together to improve the position measurements. And GPS devices have one additional advantage that makes their use attractive. They can often be programmed to give distance and heading from some fixed location. If that location is the radar, the GPS device will automatically give the range and bearing from the radar to each target used.

Azimuth and elevation angles

By now you should already have some ideas about how to check azimuth and elevation angles. The two best are using the sun and using ground targets. Both are useful; both should be used.

The sun is more useful for checking elevation angle than are ground targets, however, because of refraction near the ground. Whenever the radar antenna elevation angle is below about 10° or so, refraction is likely a problem. One advantage the sun has over ground targets is that it is higher than 10° most of the day. Further, it moves from being east of the radar in the morning to being west of the radar during the afternoon, giving the opportunity to use a single target to check azimuth and elevation angles in more than one direction.

Of course, the level of an antenna pedestal should be checked using a standard contractors level. Similar devices can be used to check the calibration of the elevation angle readout of the antenna. This should be done at several elevation angles to insure that there are no systematic biases in the angles.

Doppler velocity

This is a toughie. Actually, maybe it isn't. Frequency measurements are now quite simple to perform, especially with modern data processors. So it should not be difficult to get good velocity data from a Doppler radar. If it is correct at one velocity, it should be correct at all other velocities.

Nevertheless, it is worth verifying occasionally that speeds are correct. A simple way to do this is to compare the displayed wind speeds for the boundary layer with those of standard anemometers. There should be reasonable agreement between them. Unless there are dramatic differences present, it is probably safe to assume that the Doppler radar radial velocities are correct.

Color Figures

General comments

Most of the color figures that follow were collected by the University of North Dakota (UND) C-band Doppler weather radar. The radar was located at different places during different years. The following table gives the locations of the UND radar by year.

Year	Location	Latitude (° ' ")	Longitude (° ' ")	Altitude (ft MSL)
1988	Denver, CO	39 52 42	104 46 18	5240
1989	Kansas City, MO	39 12 38	94 36 24	1066
1990	Orlando, FL	28 25 27	81 11 29	140
1991	South Roggen, CO	40 05 42	104 20 17	4842
1992	Grand Forks, ND	47 55 17	97 05 07	895
1994+	Grand Forks, ND	47 55 18	97 05 12	920

The figures show both radar reflectivity factor Z (dBZ) and radial Doppler velocity (m/s) for a specific time and set of operational conditions. Each frame contains some housekeeping data. The following discussion describes the meaning of the various information contained in the housekeeping data.

There are color bars for each of the fields. Reflectivity colors typically start at -10 dBZ and increase in steps of 6 or 8 dB, depending upon the operator's choice. The values shown for reflectivity color bars are the minimum needed to be displayed. Thus, if green is labeled "2" and yellow is labeled "6", any echo showing green would be at least 2 dBZ but less than 6 dBZ. It is sometimes possible to interpolate between values by estimating how far above the threshold a particular echo is. The starting reflectivity and the interval are user selectable, so it is possible to set these so reflectivities can be measured to the nearest dB or better (but at the cost of a reduced range of values).

The velocity color bar covers a range of values from minus the Nyquist velocity to plus the Nyquist velocity (e.g., from -14.9 m/s to + 14.9 m/s when the PRF is 1100 Hz). There are always 9 velocity colors, so the interval shown also depends upon the PRF. For example, colors are labeled at 3.3 m/s intervals when the PRF is 1100 Hz; the highest labeled value is 13.2 m/s which is the center of the color bar, not the upper limit. Each velocity color bar is centered at the value shown. Colors to the right of zero indicate motion coming toward the radar (greens and blues); colors to the left indicate motion away from the radar (yellows and reds). It is not possible to change the velocity color bar interval on the UND radar, so it is sometimes difficult to get velocities to much better than the nearest color bar value. Of course, the exact values are recorded in the archived data and can be examined after-the-fact quite easily; it is only the real-time display that has the coarse velocity resolution limitation (reflectivities are also recorded to better precision than what is displayed).

Other information in the housekeeping field are the date (day, month, and year); time (Universal Coordinated Time, UTC [this abbreviation comes from French not English!]); some figures have two times, "V Time" and "T Time". "V time" is the time the volume scan started; volume scans typically last from 1 to 10 min, depending upon what phenomena is being studied; a single volume scan usually contains several individual sector scans at different tilts. "T time" is the time the displayed tilt began; tilts take from a few seconds to perhaps 30 s to complete, depending upon how much angular resolution is needed, how large of a sector is being scanned, and how much averaging is being done in the data processor for each recorded radial.

"Map ID" is a scan naming scheme attached to the data that records the year, month, day, "storm" number, and "map" number; "storm" is the number of the file containing specifications for the volume scan, and "map" is the number of the volume scan in the series that has been recorded as of the current time. Once a "storm" has been selected, it is usually run many times in a row, each one of which would be a new "map". "Storm" and "map" are useful for finding the data once it has been recorded on disk or on tape, i.e., for playing back the data or doing analysis of the data in post processing.

"Elevation" is the angle of the beam measured from the horizontal in degrees. The PPI figures contain a list of elevation angles to be scanned in the "storm" being recorded; these are located just below the color bar on the velocity side of the image. By knowing what angle the radar is currently displaying and looking at this list, the operator can anticipate future scans and make better operational decisions.

" PRF " is the pulse repetition rate in Hz (or pulses per second). The choice of PRF determines both the maximum unambiguous range and the maximum unambiguous velocity that the radar can resolve. See the discussion of the Doppler dilemma in Chapter 6 for the trade-offs involved in this decision.

"Sample size" is the number of consecutive transmitted pulses that are averaged together to form a single radial (or ray) of data; this is typically 64, but it can be anything from 32 to 256. The greater the number of samples averaged together, the smoother the data appear (because noise is averaged out). However, averaging takes time, so the more averaging that is used, the slower the antenna must go to maintain a specified azimuthal resolution (typically, the azimuthal separation requested is 1°, but this can be chosen to meet the requirements of the situation or the desires of the operator).

"SQI" is the value of the "signal quality index" used to collect the data. SQI ranges from 0 to 1.0 and is a threshold used to insure that the velocity data meet some predetermined tests. It helps eliminate noisy data or data of questionable quality. SQI only affects the velocity data. By setting SQI to 0.0, velocity and spectrum width data will be collected at every range gate. A typical value of SQI is 0.2 to 0.3.

"CLT" is an integer number which indicates which of several different clutter filtering thresholds has been selected. When CLT is 0, no clutter filtering is being done; CLT is typically 8 which indicates that clutter filtering is being done.

The last line says that the radar is the UND Doppler radar and gives the "gate size." Gate size is the distance between consecutive measurements along a radial; typical values for range gate spacing is 100 to 500 m, depending upon how much resolution is desired and/or how long of a distance is needed. Since the pulse duration is usually 0.6 μs (corresponding to a pulse length of 90 m), it seldom makes sense to have gates any closer together than about 100 m.

Each color image also contains data which helps relate the radar data to the real world. PPI's contain range rings and azimuth lines. The range rings are labeled in distance from the radar (km). The azimuth lines are not labeled but are usually easily determined. North is always at the top of each PPI.

RHI figures have range marks which are straight vertical lines (also labeled in km). Height lines are horizontal and given in km. Heights are measured from the radar, i.e., they are "above ground level" or AGL heights. To get the height above sea level (a useful number when aircraft are involved in an operation), it is necessary to add the height of the radar above mean sea level (MSL). The table given earlier contains this information (albeit, in feet; the radar display gives heights in kilometers; it is necessary to convert to a consistent set of units. Pilots use feet MSL; many radar meteorologists use kilometers).

Some of the PPI figures also contain geographical or auxiliary coordinate system data. For example, Color Figures 6 and 11 contain overlays showing the runways at Stapelton International Airport (58 km range, 235° azimuth from the UND radar). Centered on Stapelton is a series of dashed circles at 5 n mi intervals; between 20 and 25 n mi is a series of heading marks, giving directions relative to magnetic north (aircraft operations are usually conducted in magnetic headings and nautical miles relative to a particular radio navigation aid; these marks allows the radar operator to tell the pilot where things are in coordinates that are useful to the flight crew). The overlay also contains symbols to indicate other things of importance to radar operations. Squares at 46 km, 327° and 45 km, 235° show the locations of the Colorado State University CHILL S-band radar and the National Center for Atmospheric Research Mile High Radar (MHR), respectively. The runways of Buckley Air National Guard are at 57 km and 217° azimuth. Triangles at 58 km, 265° and 35 km, 285°, respectively, give the locations of the Boulder

Color Figures

Atmospheric Observatory (a 300-m tall instrumented tower) and the Plattville vertical wind profiler operated by the National Oceanic and Atmospheric Administration. Knowing where these installations are during radar operations is helpful because it allows the operator to make better decisions for certain operations.

Color Figure 1

PPI of ground clutter from South Roggen, Colorado, radar site on 7 March 1991. There is virtually no clear-air return at all on this figure. The slight background echo on the reflectivity field is receiver noise. Near the radar is local ground clutter; 80 to 100 km west are the Rocky Mountains. The reflectivity of ground targets can be quite variable and quite large. Some ground targets likely saturate the radar receiver.

The velocity of all ground clutter on this figure is indicated as white (zero radial velocity). There are a few point targets in the clear region between the mountains and the nearby ground clutter which do have velocity; a corresponding reflectivity point should be detectable on the Z field. These are possibly birds or aircraft. Stapelton International Airport is located at about 58 km and 235° azimuth from the radar.

There are three or four white spikes in the velocity data pointing away from the radar to the northeast through south just beyond the nearby ground clutter. Farther out (80 to 100 km) along the same radials are some more echoes. These have weak reflectivities. These echoes are caused by the radar beam pointing in the directions indicated but the signal is reflecting off of nearby farm buildings (within 1 to 3 miles from the radar) and hitting the Rocky Mountains far to the west. These targets are always present in our low-level scans from this radar site (see Color Figure 6). These data were collected with the clutter filters turned off (CLT = 0); when clutter filtering is used, these targets are less noticeable.

Color Figure 2

This is clear-air return detected by the radar on 6 June 1989 in Kansas City, Missouri. The data were collected at 0.3° elevation angle with the clutter filters off. Notice that the ground return pattern is much different at this location than it was in Denver. Zero velocities are evident near the strong ground targets of downtown Kansas City.

Winds on this day were from the south, with near-zero velocities at the ground and increasing to about 6 m/s higher up in the boundary layer. The source of this clear-air return is likely insects (which are slow enough flyers that they make good tracers to get the wind speed). The reflectivities of the clear-air return is on the order of -10 to +6 dBZ, depending upon where the measurement is made. Notice that where the reflectivity is very weak, the velocity data is also noisier; where the reflectivities exceed -2 dBZ , the velocity data looks quite consistent and reliable.

Color Figures 3 (PPI) and 4 (RHI)

This figure shows a cold front approaching the UND radar from the northwest on 18 June 1989 at Kansas City. On the PPI the reflectivities in this storm are 54 dBZ or higher. The reflectivity structure shows stronger reflectivities near the leading edge of the storm with weaker reflectivities behind it.

Between the radar and the storm is some clear-air echo of -10 to +20 dBZ reflectivity. The velocities in the clear air are away from the radar at speeds up to 10 m/s.

Color Figures

At a range of 12 km (toward the NW) the winds abruptly change to westerlies with speeds reaching as much as 28 m/s (near 18 km range, 340° azimuth); notice that these velocities are aliased.

The RHI of this storm was taken along the 339° azimuth. The reflectivities show the cores of strongest reflectivity extend up to about 5 km AGL with the storm top reaching up to 12 to 13 km.

The RHI of velocities shows much of the echo is receding toward the northwest through most of the depth of the echo. However, the air behind the cold front is approaching the radar rapidly over a shallow layer which is only about 2 km deep. It is interesting in this and other RHI's of wind situations to notice how shallow certain wind events really are. It is difficult to recognize this from looking at PPI's only.

Color Figure 5

This figure shows a microburst detected by the UND radar in Denver on 11 July 1988. This microburst, located at 15.5 km and 200° from the radar, is very strong. It shows a velocity difference from one side to the other of 37 m/s (nearly 75 knots). This microburst would be classified as wet because the reflectivities near the center are as strong as 54 dBZ. This would correspond to a rainrate of about 89 mm/h.

This microburst was detected by an automatic microburst-detection algorithm developed by MIT Lincoln Laboratory and being run on the Lincoln FL2 S-band radar. Several aircraft diverted their landings, in part because of this warning.

Color Figure 6

This figure shows both real (first-trip) and range aliased or second-trip echoes. These data are from the UND radar at South Roggen, Colorado, on 12 March 1991. The real echo toward the west through NW near 60 km range is only of moderate intensity (near 38 dBZ) and is moving toward the radar at speeds up to 18 m/s.

Second-trip echoes occur along several azimuths, notably toward the north, NE and SSW. These echoes are moderately weak and do not show on the velocity display (see Chapter 6 for a discussion of velocity signal quality checks).

Also detectable in the nearby ground clutter toward the northeast and southeast are the reflection artifacts discussed in Color Figure 1.

Color Figures 7 (PPI) and 8 (RHI)

This is an example of a gust front detected by the UND radar at Kansas City on 27 August 1989. The gust front is best seen in the reflectivity data on the PPI where it is a curved band of green/yellow echo (up to 14 dBZ) on the east side of the storm echoes. Unfortunately, there are some second-trip echoes overlaid on the gust front (see second-trip echo discussion in Chapter 6). The range-aliased echoes are rejected by the data quality checks in the velocity data.

The gust front is harder to see on the velocity display. There is some clear-air return northwest of the radar moving toward the NW. The main mass of echo is moving toward the radar. Between these, at the same position as the thin-line echo on the reflectivity display, is a faint band of white and blue pixels. These mark the boundary of the approaching air.

The RHI data also contain the gust front data, but it is very difficult to see. The storm producing the gust front is clearly visible. It has a well-defined anvil echo flowing rapidly toward the radar. It also has a strong reflectivity core close to the leading edge of the main echo wall. The gust front is located near 14 km range (on this 309° radial). It is obviously quite shallow.

One interesting feature is that the surface air moving toward the storm rises up above the gust front and enters the main cell between 2 and 4 km, extending back into the storm as far as 38 km range. This is the main inflow feeding the storm with moisture. Some of this air turns upward and comes back toward the radar in the anvil while another part of it turns downward and becomes part of the forward-moving gust-front air.

Color Figures 9 (PPI) and 10 (RHI)

This is an example of a "bow" echo that was detected by the UND radar in Kansas City on 25 May 1989. There is a narrow band of strong echo ($Z > 54$ dBZ) located as close as 25 km straight west. Just in front of this high-reflectivity core is a band of very strong eastward moving winds, with speeds up to 32 m/s. The name "bow" echo is related to the shape of the reflectivity pattern of the echo and how the strong winds emanate from it (Fujita, 1985).

The RHI through this echo was made 4 min 15 s before the PPI along an azimuth of 276°. The band of strong winds is seen as a very shallow region of approaching velocities centered near 28.5 km. During the 4.25 min period between the RHI and the PPI, the core of the strong winds moved 8 km, giving a speed of 31.7 m/s, in good agreement with the Doppler velocity shown by the radar.

One thing is evident in the velocity RHI data. The data are aliased at least twice. The strongest approaching velocity appears to be near 9 km altitude at 20 km range and is on the order of 55 m/s (107 knots!)

Color Figure 11

This figure contains several interesting features, the most significant of which is a tornadic vortex signature (TVS) located near 35 km range at 215° azimuth. These data were collected by the UND radar at Kansas City on 25 May 1989. The reflectivity of this storm is apparently well above 54 dBZ. In the vicinity of the TVS on the southwest edge of the storm is an appendage which resembles a "hook echo," one of the classic indicators of a tornado.

This figure also shows the effects of attenuation behind the strong echo. The clear air return which is detectable around the echo does not show on the far side of the echo. There is also a hint of attenuation on the far side of the main echo core. It is impossible to know exactly how much attenuation is taking place, but it appears to be on the order of 10 to 15 dB or more.

Another feature of this storm is the approaching wind shift to the northwest. The zero isodop changes direction rather abruptly near 300° azimuth at 20 km range. Winds are from the southwest over most of the region but appear to be from the northwest behind this wind-shift line.

Color Figure 12

This is a second example of a tornadic vortex signature. This one is from the UND radar on 15 June 1988 near Denver. The TVS is located near 19.5 km at 220° azimuth. The velocity field shows an abrupt change in speed over a distance as small as one or two radials. The velocity change is azimuthal in nature, not radial. That is, in moving along in azimuth at a constant range, we see approaching velocities on the left (as viewed from the radar) with receding velocities to the right. The reflectivity field in this area is again quite strong, being well in excess of 54 dBZ. A hook-like appendage extends away from the main echo at this point also. There was a tornado confirmed at this location near this time.

A second region of strong rotation is visible on this figure just beyond 29 km range at almost exactly the same azimuth. This one is somewhat larger in size, but, because of the 2.2° tilt at which the data were collected, it represents data higher in the storm (near 1.2 km AGL). This could be a tornado mesocyclone. The reflectivity of this storm is also over 54 dBZ.

Color Figures 13 and 14

These are examples of backing and veering winds detected by the UND radar at South Roggen, Colorado. Both cases cover a range out to 70 km, and both show similar reflectivities (near 26 dBZ).

On the 25 January 1991 case, notice the backward "S" shape of the zero isodop; this is the Doppler velocity signature of backing wind situations. The winds at the surface are light, about 3.3 m/s, from the NE. Winds increase to 13.2 m/s at a range of about 8 km (460 m AGL), still from the NE. Above about 900 m (17 km range), the winds begin to back. This continues to a height of about 1200 m (21 km range). The strongest winds occur near the top of the echo. Winds here are from the WNW at a speed of 18.5 m/s (1500 m AGL).

The 29 January case has an "S"-shaped zero isodop. In this case the surface winds are again from the northeast but at a speed near 9.9 m/s. Winds increase with altitude to a maximum of 13.2 m/s at an altitude of 400 m AGL. Above this the winds begin to veer, reaching a maximum of 18.5 m/s at an altitude of 1400 m AGL from the SSE. Above this the winds begin to decrease somewhat but then begin increasing again, reaching a maximum of 21.8 m/s from the west at an altitude of 3.9 km (60 km at 270° azimuth).

One minor but interesting echo on this figure is a small region of zero velocity echo straight north at 42 km. This is a sidelobe echo from a very strong ground target located at the surface at this range. Notice that this echo does not show in the reflectivity data.

Color Figure 15

Example of the NEXRAD storm-track algorithm produced by the WSR-88D radar located near Norman, Oklahoma. This figure shows the radar reflectivity data for storms at 2108 UTC on 21 March 1991 and were collected at an elevation angle of 0.5°. The "housekeeping" data along the right side of the figure tell where, when, and how the data were collected. The map overlay shows the county boundaries around the radar

The figure also shows the position of these storm cells over the past hour (solid, black dots generally southwest of the current positions) and the forecast position at 15-min intervals into the future (plus signs on the extrapolated tracks). These results came from the storm track algorithm.

Color Figure 16

This figure shows the output of the storm total precipitation algorithm. The color bar at the right shows that some of the storms moving across this area produced precipitation totals as large as 4 inches. Notice that this display covers exactly the same area as that of Color Fig. 15 and for the same storm. However, Color Fig. 16 is for the period from 0533 UTC through 2204 UTC whereas Color Fig. 15 is for a time of 2108 UTC.

Notice that individual storm tracks are recognizable on this figure. The storms shown on Color Fig. 15 were moving toward the ENE, as shown by their tracks. Other storms moved somewhat more toward the northeast. These likely occurred before those of Color Fig. 15.

Color Figures 17 and 18

These figures show the radar reflectivity factor (2344 UTC, Fig. 17) and Doppler radial velocity (2344 UTC, Fig. 18) for thunderstorms on 5 June 1996 at 0009 UTC from the Mayville, North Dakota, WSR-88D. Each image has a color scale to give reflectivity and velocity. The reflectivity image also shows storm positions and forecast positions similar to those on Color Fig. 15. The table at the top gives storm identification numbers, the azimuth and range to the center of the storm, the forecast movement, track errors, and the reflectivity and height for several cells identified by the computer software algorithms as individual storms.

On the velocity image colors in the greens are approaching the radar while those in the reds are going away from the radar. The general storm movement is from the northwest toward the southeast.

Color Figures

Color Figure 19

Hurricane Danny detected by the Mobile, Alabama, WSR-88D radar as the hurricane came on shore on 19 July 1997. This was a slow-moving hurricane that produced over 20 in of precipitation (as measured by radar).

Color Figure 20

Bird echoes detected by the Nashville, Tennessee, WSR-88D radar at 1108 UTC on 12 August 1995. Each of the ring or donut echoes is centered on a roosting site of the birds that are being detected by the radar. See Chapter 11 for more about this figure.

Color Figure 21

Mosaic of 30 radars at a variety of times from 1995 (Nashville only) and July and August 1997. The images are all near sunrise. In fact, several of the radars show a few radials of signal in the direction of the rising sun. Most of the radars were operating in the "clear-air" mode so they show reflectivities from –20 to +28 dBZ in 4-dB steps. These radars generally have red colors in their images. Some of the radars were operating in the "precipitation" mode so reflectivities go from 5 dBZ to 75 dBZ in 5-dB steps.

There are approximately 50 bird rings showing on this mosaic. The ring echoes vary in size for at least two reasons. One is how long before or after sunrise the image was collected. The second is how many birds were roosting at the particular site. See also Chapter 11 for additional discussion of bird ring echoes.

Some of the earliest bird echoes discussed in the literature (e.g., Elder, 1957) show that bird ring echoes expand with time. Occasionally, others have reported that multiple rings emanate from the same location, with waves of birds leaving at different times.

Color Fig. 21 may show at least one ring with multiple waves of birds. South of the Oklahoma City radar is a small ring echo with a larger ring around it (at least toward the east).

One reason the bird rings show so well is because of refraction. At many locations there is a nocturnal inversion in the atmosphere caused by radiational cooling on clear nights. Surface temperatures are cool with warmer temperatures aloft. This causes the radar beam to bend down much more than under standard propagation conditions. Birds generally fly relatively close to the ground. Birds that are going out to forage for the day probably don't get much higher than a few hundred feet. Figure 3.6 shows that a radar beam starting at 0.5° would be as high as 5 km at a range of 230 km from a radar, the maximum range for the NEXRAD images used in the mosaic. Starlings or other social birds would not likely be at altitudes as high as 16,000 ft! If it were not for anomalous beam bending in the early morning hours near sunrise, the radar beam would pass well above the birds and never detect them. So seeing these bird rings takes a combination of things: good radar sensitivity, superrefraction, and lots of birds! It also takes getting up early enough to see them before the birds disperse and land and before the beam rises to its "standard" height as the day warms up.

Color Figures 19-21 are based on NEXRAD images provided courtesy of INTELLICAST (http://www. intellicast.com/). INTELLICAST is a registered trademark of WSI Corporation.

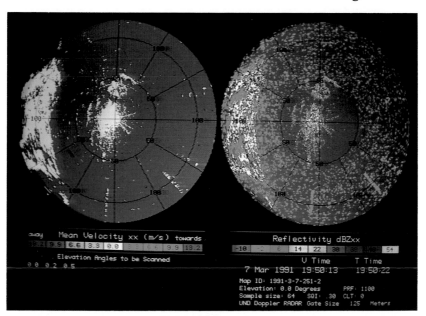

Color Figure 1

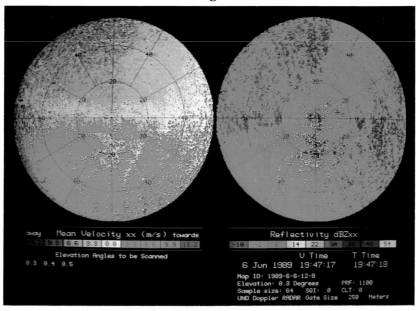

Color Figure 2

Color Figures

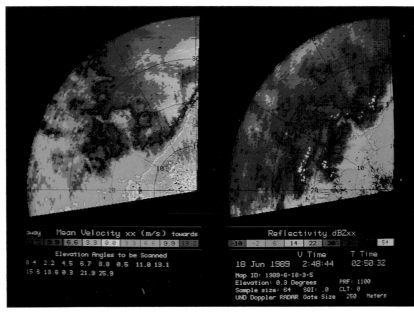

Color Figure 3

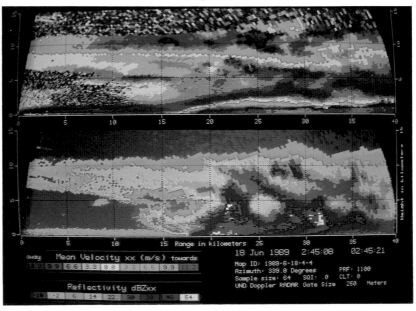

Color Figure 4

Color Figures

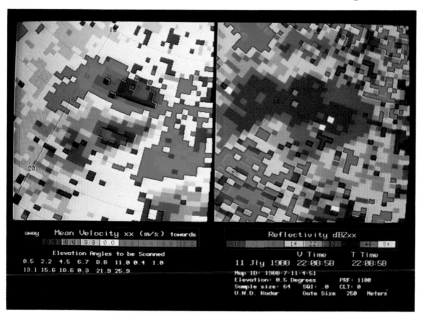

Color Figure 5

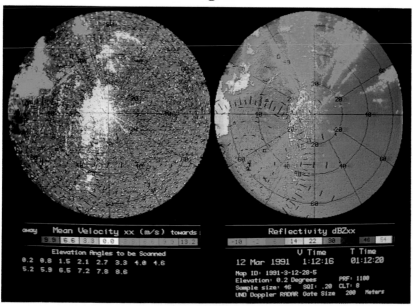

Color Figure 6

Color Figures

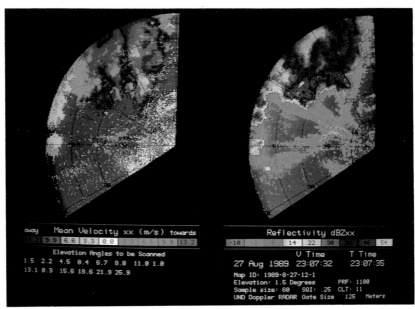

Color Figure 7

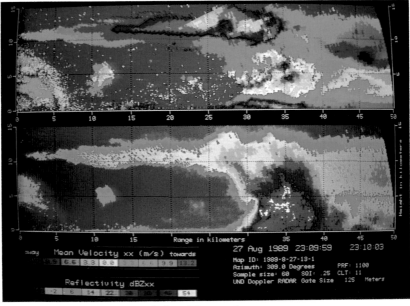

Color Figure 8

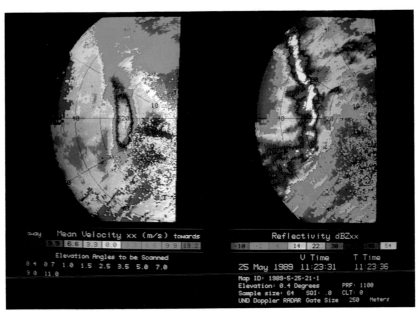

Color Figure 9

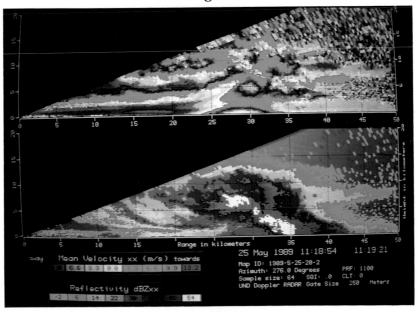

Color Figure 10

Color Figures

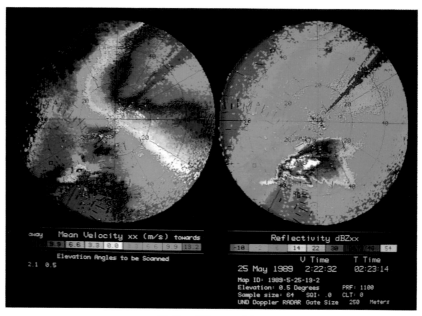

Color Figure 11

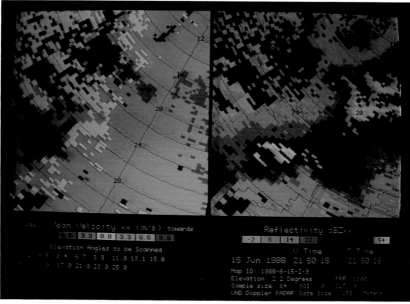

Color Figure 12

Color Figures

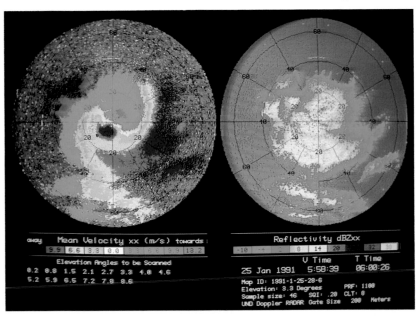

Color Figure 13

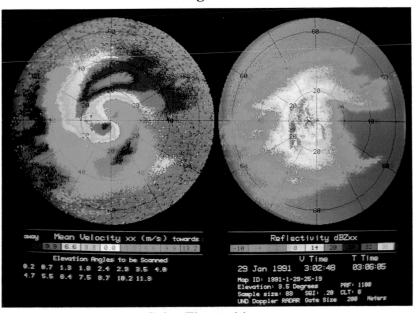

Color Figure 14

Color Figures

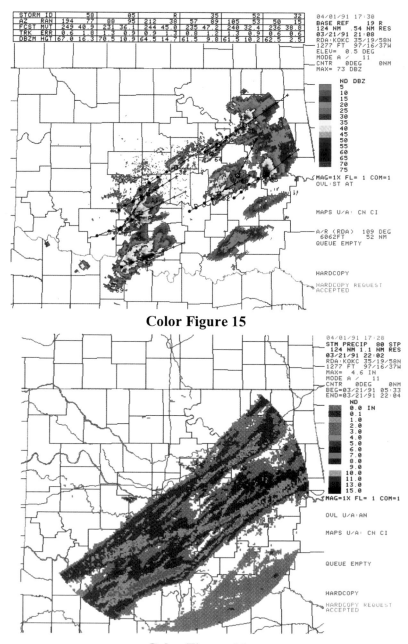

Color Figure 15

Color Figure 16

Color Figures

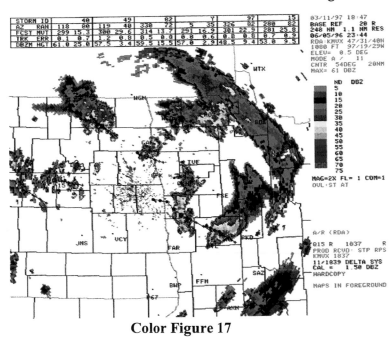

Color Figure 17

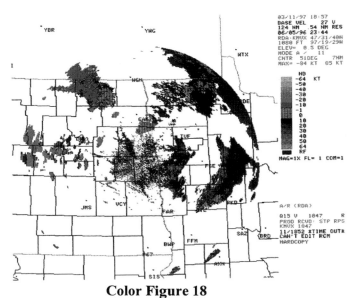

Color Figure 18

Color Figures

Color Figure 19

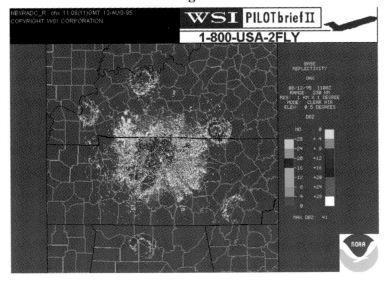

Color Figure 20

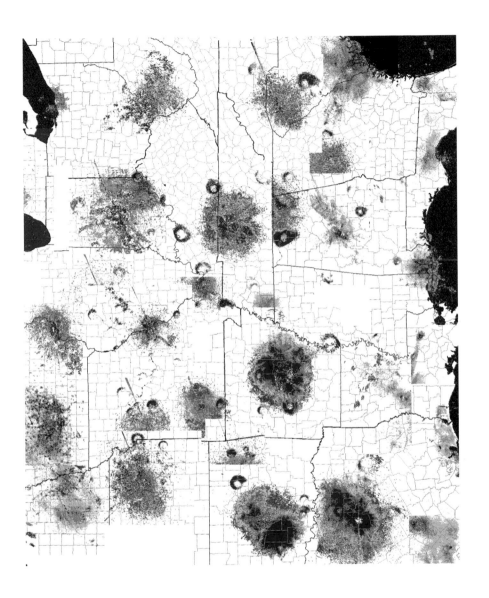

Color Figure 21

R. E. Rinehart and Marshal Hagen, 1997

Appendix A

Logarithmic Units

Most measurements have four components associated with them: The parameter being measured, its magnitude, its units and the symbol we attach to it. As an example, rainrate is given the symbol R and is usually measured in units of mm/h or in/h. The magnitude might be 5 or 50 or 500, depending upon how hard it is raining. There are many possible parameters which can be measured. Unfortunately, there is a much more limited number of symbols we can attach to them. Using the English language, for example, we can have 52 letters (a, A, b, B, etc.). Of course, we can combine these, but then we loose some of the advantage of having short symbols. We can use letters from other languages. The Greek alphabet has been extensively used in the sciences. Other languages are also sometimes used, including Russian, Hebrew, and some of the European languages (Å, for example, has been used for the Ångström, a unit of length $= 10^{-10}$ m).

The number of units is also fairly large, but for a given parameter only a few possibilities usually exist. There are actually only a handful of "dimensions." These include length, mass, time, charge, and temperature. Length, however, can be expressed in units of: meters, feet, inches, miles, nautical miles, furlongs, rods, light years, etc., etc., etc. By using metric units we can simplify our life considerably. Prefixes such as milli-, deci-, kilo-, mega-, etc., can be added to a unit to modify the magnitude of the measurement.

One of the major purposes of all units is to make our measurements a convenient size. I have the idea that the human mind cannot comprehend numbers larger or smaller than a certain size. Numbers larger than perhaps a few hundred begin to lose their meaning. Numbers smaller than perhaps 1/100 do too. If numbers get too large, we talk about thousands of dollars instead of dollars, miles instead of inches or feet, and so forth. We could, for example, measure distances between cities in inches, but that would be exceedingly inconvenient, not to mention incomprehensible.

The same problems exist in radar meteorology as in other human endeavors. In order to understand the size of things, we have to have measurements which are limited to a reasonable range of values. Radar reflectivity, as mentioned before, ranges from exceedingly small values to incomprehensibly large values. By converting to logarithmic units, we compress reflectivities into a range of from about -20 or -30 dBZ to more than 70 dBZ. And usually quoting reflectivities to the nearest whole decibel is good enough. If more accuracy is needed, we can report our values to the nearest 0.1 dB. Reflectivity measurements more precise than this are almost never justified.

Because we work with logarithmic units so frequently in dealing with radar, we need to be able to convert quickly from linear units to logarithmic units and vice versa. So let's review the most commonly used logarithmic units and how to convert back and forth.

When we take the ratio of two powers, we get the power ratio given by

$$ratio = \frac{P_1}{P_2}$$

(A.1)

a unitless number [if the two powers are measured in different power units, we must convert to a common power unit in order to get the units to cancel]. Because of the large range that this ratio can take in electrical measurements, the logarithmic power ratio was defined as

$$logaithmic\ ratio = 10\ log_{10}\left(\frac{P_1}{P_2}\right)$$

(A.2)

and is measured in units of decibels, abbreviated dB.

In these and other units, certain conventions are usually followed. Units named after people are capitalized. Decibels are named in honor of Alexander Graham Bell; hence, the capital B. The "d" stands for the standard metric prefix "deci-" which means $1/10^{th}$ (1 bel = 10 decibel). Incidentally, the capital used with the symbol for a unit is not used when the unit is spelled out, as in "bel". Other units named after people are the hertz (1 Hz = 1 cycle/s), coulomb (a unit of electrical charge), farad (unit of electrical capacitance, named after Faraday), watt (unit of electrical energy), and kelvins (degrees from absolute zero temperature). Note that Celsius and Fahrenheit are always capitalized; all of the others are not.

Now, returning to power measurements, if we want to express absolute powers on a linear scale, we use units of μW, mW, W, kW, MW, etc. Again, the

Appendix A

range of possible powers is large enough that it is convenient to compress the range by using logarithmic units. In this case we can define logarithmic powers as

$$P = 10 \, log_{10}\left(\frac{p}{1 \, mW}\right) \tag{A.3}$$

expressing the logarithmic parameter in units of dBm, where the "m" means decibels relative to a power of one milliwatt (i.e., 1 mW). Any time we see a power in units of dBm, we know it is an absolute power; any time we see a parameter in units of dB, we know we are dealing with a power ratio or a relative measurement.

Let us review some basic mathematics used in our manipulations. When we multiply or divide one number by another, we write

$$y = ax$$

or

$$z = \frac{u}{b}.$$

If we take the logarithms of both sides of either of these, we can write

$$log(y) = log(ax)$$

$$= log(a) + log(x),$$

and

$$log(z) = log\left(\frac{u}{b}\right)$$

$$= log(u) - log(b).$$

If we want to work with decibels instead, we would simply multiply all terms by 10. The important point to notice in these expressions is that the product of two numbers on a linear scale is the sum of the logarithms of each number. Division of linear parameters is given as the difference in the logarithmic parameters. Going the other way, any time we see the sum or difference in two logarithmic parameters, it really means that the linear parameters are being multiplied or divided.

One of the great advantages of using logarithmic parameters is that multiplication and division are reduced to simple addition and subtraction. In fact, in the days before hand calculators and computers, it was much easier to add and subtract than to multiply and divide, especially for very large or very small numbers. This is perhaps another reason why logarithmic units were used so extensively by those dealing with powers and their ratios.

The following rules can be applied to working with logarithmic units.

$$dBm \pm dB \rightarrow dBm$$

$$dBm - dBm \rightarrow dB$$

$$dBm + dBm \rightarrow [Not\ used]$$

Appendix A

$$dB \pm dB \rightarrow dB$$

In the first case, addition or subtraction of a power ratio (the "dB" term) with a term in "dBm" always gives "dBm" units. This corresponds to multiplying or dividing an absolute power by a unitless numerical ratio or parameter. The product has the same units as the first parameter.

In the second case, the difference in two absolute powers measured in logarithmic units corresponds to the division of one power by another power. The units of these two linear powers would cancel, yielding a unitless ratio; hence, on a logarithmic scale, it should have units of decibels.

The third case, the sum of two logarithmic absolute powers, while mathematically possible, is not particularly useful. It would correspond to multiplying two powers together, giving units of power squared. I cannot think of any particular use of this manipulation.

Finally, the sum or difference of two logarithmic power ratios is another power ratio. This is the same as multiplying or dividing two linear power ratios to get their product or quotient.

Exactly the same set of manipulations can be done with logarithmic and linear reflectivity units. That is, we can use:

$$dBZ \pm dB \rightarrow dBZ$$

$$dBZ - dBZ \rightarrow dB$$

and $dBZ + dBZ$ is again not used.

In working with logarithmic units of reflectivity, we can say, for example, that a 50 dBZ storm has increased 5 dB to become a 55 dBZ storm. Or a 50 dBZ storm has decreased 10 dB to become a 40 dBZ storm. The difference in intensity between a 50 dBZ storm and a 45 dBZ storm is 5 dB.

Regarding reflectivity units, especially the logarithmic units, there exists a lot of incorrect usage in the radar meteorological community. It is not unusual to hear or see someone talking about the "dBZ's of a storm was 55." This is patently wrong! We never talk about the "kilometers of a storm was 10" when what we really mean is the "height of the storm was 10 km." Nevertheless, this is frequently done with reflectivity units. The correct terminology is, as with all measured quantities, to talk about the parameter or to substitute the correct symbol in its place. So, we should say "the reflectivity of the storm is 55 dBZ" or whatever value is correct.

In addition to power and reflectivity, there are a number of other parameters which are frequently measured in logarithmic units. Among these are antenna gain, waveguide losses, atmospheric losses, attenuation of radar signal from rain, and a few others.

Another parameter conveniently measured on a logarithmic scale is backscattering cross-sectional area σ. σ varies over as many orders of magnitude as does radar reflectivity factor Z. For example, σ from a single cloud droplet, raindrop, or small insect might be as small as 0.001 cm^2. σ increases for birds and small aircraft to values of perhaps 1 to 1000 cm^2. Large aircraft can have backscattering cross-sectional areas as large as 100000 cm^2. Buildings can have σ's as large as 10^8 cm^2 or larger. By defining

$$\Sigma = 10 \log\left(\frac{\sigma}{1\ cm^2}\right) \qquad (A.4)$$

and measuring Σ in units of dBσ, we compress these values to a range of -30 to 80 dBσ, a much more manageable set of values.

There is another logarithmic unit also being used for backscattering cross-sectional area σ, and this is to give the reference value in m^2 rather than in cm^2. The logarithmic units for this is dBsm for "decibels relative to an area of 1 m^2". The advantage of referencing areas to 1 cm^2 to convert into logarithmic units is that the magnitudes of the resultant measurements are very similar to those of radar reflectivities in dBZ, i.e., most are between about 0 and 60 dBσ. When the reference value is 1 m^2, logarithmic backscattering cross-sectional areas range from about -40 to +20 dBsm, a somewhat more awkward range of values for meteorological use.

One final problem that seems to be associated with logarithmic units that never occurred with linear units is the use of the subscripts. The subscript "e" is used to designate "effective" or "equivalent" radar reflectivity factor Z_e. We use Z_e in place of Z when the reflectivities which are being measured are from storms where it is not known if the detected particles meet the Rayleigh condition or not or when the measurements are being made by radar. However, the correct place to put the subscript is on the parameter's symbol, not on the units. Subscripted variables are very common; subscripted units are not. We never refer to the size of a storm as being 12 km_h, to indicate "height." It never occurred to anyone to try to add subscripts to mm^6/m^3; it is equally incorrect to add them to logarithmic units.

Life on a Logarithmic Scale:

If you work frequently with radar data, you will find it necessary to convert from linear to logarithmic units. Since many calculations are just done to get a reasonable guess as to an answer, we can often live with answers that are within a decibel or so most of the time. By knowing how decibels are calculated and memorizing a couple of numbers, we can come up with a way to calculate back and forth from linear to logarithmic units.

First of all, decibels are calculated using logarithms to the base ten. The logarithm of 1 (to any base) is zero; to the base ten, the logarithm of 10 is 1; the log of 100 is 2, etc. Converting to decibels gives the following:

Linear value	logarithm	decibels
0.001	-3	-30
0.01	-2	-20
0.1	-1	-10
1	0	0
10	1	10
100	2	20
1000	3	30
etc.		

So, using the above numbers as examples, we should be able to handle any power of ten quite easily.

Now we need to handle numbers between 1 and 10. One of the numbers you will have to memorize is that a factor of 2 is 3 dB. That is, $\log(2) = 0.30103$ so 10 $\log(2) \cong 3.0$ dB. When we multiply two numbers on a linear basis, we determine their product using logarithms by adding their logarithms. Thus, $2 \cdot 2 = 4$, so $\log(2) + \log(2) = \log(4)$. Using decibels, we calculate the decibel equivalent of 4 from 3 dB + 3 dB = 6 dB.

313

Following that logic, the decibel equivalent of 8 is thus 9 dB. So, we can write the following equivalents:

Linear value	logarithm	decibels
1	0	0
		1
		2
2	0.3	3
		4
		5
4	0.6	6
		7
		8
8	0.9	9
10	1.0	10

We can use the factor-of-two trick three more times. 10 divided by 2 = 5; 5/2 = 2.5; and 2.5/2 = 1.25. So, the decibel equivalent of 5 should be 3 dB less than the decibel equivalent of 10 or 7 dB. Similarly, 2.5 is 3 dB less than 5, or 4 dB; and 1.25 is 3 dB less than 2.5 or 1 dB. Now we can add to our table, giving

Linear value	logarithm	decibels
1	0	0
1.25	0.1	1
		2
2	0.3	3
2.5	0.4	4
		5
4	0.6	6
5	0.5	7
		8
8	0.9	9
10	1.0	10

There are still three missing values in our table. To fill in one of these we need to memorize one other number, the linear value which has a decibel value of 5 dB; this number is 3.162277..., or almost π or almost 3.2 or almost 3; for rough calculations, these are all the same. So, by remembering that the decibel equivalent of π or 3 ≅ 5 dB, we can fill in the rest of the table. Twice π is 6.4 or approximately 6, so the decibel equivalent of 6 is 3 dB more than that of π or 8 dB. And half π is 1.6 or about 1.5, so its decibel value is 2 dB. So, now we can complete our table:

Linear value	logarithm	decibels
1	0	0
1.25	0.1	1
1.5	0.2	2
2	0.3	3
2.5	0.4	4
3	0.5	5
4	0.6	6
5	0.7	7
6	0.8	8
8	0.9	9
10	1.0	10

Notice that the values in this table have been rounded somewhat to make them easier to remember. For quick calculations, they are perfectly fine.

Since not all numbers are between 1 and 10, a slight complication arises when larger or smaller numbers are encountered. For example, if we need to convert a number like 17,000 into a decibel value, we need to recognize that this number can be written as $1.7 \cdot 10^4$. Then we know that the decibel value will be 10 log(1.7) + 10 log (10^4) ≅ 2 dB + 40 dB = 42 dB. My calculator gives the precise answer as 42.3 dB, but the approximation is fine for top-of-the-head calculations.

Numbers smaller than 1 are not much more difficult. Take the number 0.00678. This is approximately $6.8 \cdot 10^{-3}$, giving 8 dB + (-30 dB) = -22 dB.

Working from decibels to linear values should be equally simple. Take a number like 44 dBm. Here we can again separate this into 4 dB + 40 dB, giving $2.5 \cdot 10^4$.

Negative decibel values are slightly more difficult. For example, -83 dB = -90 dB + 7 dB. This manipulation separates the power-of-ten term (-90 dB) from the numerical factor (7 dB). This then becomes $5 \cdot 10^{-9}$.

All of these calculations work essentially the same for units of dB, dBm or dBZ. The only difference is that dBm and dBZ are absolute units of measure, so the linear equivalents of them must have units attached. You can verify this by working backwards from the definitions of these given earlier. For example, hail is often associated with reflectivities greater than 55 dBZ. What is the linear equivalent of this? Again, we can write this as (5 + 50) dBZ which gives $3 \cdot 10^5$ mm^6/m^3.

With a little practice, you should be able to convert any linear number into its decimal equivalent in your head and have your answer correct to within a decibel. Similarly, you should be able to convert decibel values into linear values and be correct to within $\pm 25\%$ or better.

Appendix B

Error Analysis

In the "Preface" to his excellent book on error analysis, John Taylor (1982) begins:

> All measurements, however careful and scientific, are subject to some uncertainties. Error analysis is the study and evaluation of these uncertainties, its two main functions being to allow the scientist to estimate how large his uncertainties are, and to help him to reduce them when necessary. The analysis of uncertainties, or "errors," is a vital part of any scientific experiment, and error analysis is therefore an important part of any...experimental science. It can also be one of the most interesting parts.... The challenges of estimating uncertainties and of reducing them to a level that allows a proper conclusion to be drawn can turn a dull and routine set of measurements into a truly interesting exercise.

Radar meteorology is certainly an experimental science; meteorologists make frequent measurements using radar data but often have little idea of the uncertainty of their measurements. Each measurement contains some uncertainty. Rather than simply trusting our calculators or computers to give us precise answers, we need to be able to determine the uncertainty in our measurements based on the uncertainties of each of the components of the parameter being used. In the discussion that follows, we will examine the use of error analysis as it applies to both linear and logarithmic measurements.

We need to differentiate between "error" and "mistake." Scientific measurements almost invariably contain small errors or uncertainties in them. These are unavoidable and are an accepted part of making measurements. "Mistakes", on the other hand, are avoidable and should not be included in the error analysis which is done as part of an experiment. For instance, if someone reads a meter incorrectly or reads it correctly but writes the wrong number down, that is a mistake; as soon as it is discovered it should be corrected, repeating the experiment if necessary.

Errors in our measurements, however, depend upon the kind of instrument used and how the measurements are made. Measuring a person's height, for example, depends on the quality of the ruler used and how the measurements are made. A ruler calibrated with reference (tick) marks every 1 in would make it possible to get the height only to the nearest inch unless the person making the measurement could "interpolate" or estimate the actual height between two tick marks. Even then, this could improve the precision of the measurement only to the nearest 1/5 or 1/10 in. If the ruler had tick marks every 1/16 in, the height measurement could probably be made more precisely.

At this point it is important to distinguish between two terms which are often confused: accuracy and precision. Accuracy is defined as freedom from mistake or error, conformity to truth or a standard, degree of conformity of a measure to a standard or a true value. Precision is defined as the degree of refinement with which an operation is performed or a measurement stated. Accuracy relates to how close a measurement is to the truth while precision relates, for example, to the number of digits we use to represent the measurement. In the case of measuring heights, if the ruler used had been broken off at one end and was actually shorter than it should be, the measurements made with it would not be very accurate (unless the missing distance was discovered and added back into the measurement). We could still read the ruler with the same precision. It is certainly possible to have very precise measurements which are not very accurate. This is exactly what happens when we make measurements that have uncertainties of a few percent but give answers from a calculator, for example, which gives seven or eight or ten digits. A friend who is an engineer once quoted a saying used in his office which relates to construction projects: "Measure it with a micrometer, mark it with a chalk, and cut it with an ax."

In general we should give no more digits in stating a single measurement than can be justified by the instrument used. A careful reading of a scale can usually yield measurements which are a fraction of the smallest division on the scale, but sometimes the best we can do is read answers to the nearest division. That is all the accuracy that should be recorded.

When we see numbers written down we often interpret the stated value to being precise to plus or minus half the smallest digit given. For example, if a length is quoted as being 17 mm, we usually interpret this to mean that the length is somewhere between 16.5 and 17.5 mm; otherwise it would have been reported as 16 or 18 mm instead of 17 mm. This is often a correct interpretation. It is better, however, to state the uncertainty explicitly. It is possible that the real length is 17.00 ± 0.01 mm and the reported answer was given only to two digits because trailing zeros were dropped from the answer (computers often drop trailing zeros, for example; alternatively, they can also add unjustifiable trailing zeros sometimes). By stating the uncertainty explicitly, there can be no doubt about the precision of the value.

How much precision is justifiable in stating uncertainty? A measurement of 133.451 ± 0.322 in is really not known to three decimal places. About all we can justify for this measurement is that it is 133.5 ± 0.3 in. For computational purposes, we might want to carry one more digit, but we should round our answers so that we have uncertainties of one or two digits; one digit uncertainties are adequate most of the time.

When we make measurements, then, we should do two things: Determine our best estimate of the measurement, and determine our estimate of the uncertainty of the measurement. Another way of stating this latter step is to determine the probable range of values within which the best estimate lies. Usually our best estimate will be centered within this range. Any measurement can be stated in the following general way [Note: Most of the following notation and rules come from Taylor]:

$$(measured\ value\ of\ x) = x_{best} \pm \delta x \qquad (B.1)$$

where δx is the uncertainty of the measurement. Notice that δx must have the same units as x_{best}.

Rule for Stating Uncertainties: In making general measurements, uncertainties should usually be rounded to one significant figure.

Rule for Stating Answers: The last significant figure in any stated answer should usually be of the same order of magnitude (i.e., in the same decimal position) as the uncertainty.

However, numbers used in calculations should be kept with one more significant figure than is finally justified. This will reduce the inaccuracies introduced by rounding the numbers. At the end of the calculation the final answer should be rounded to remove this extra figure. There is an exception to the rule stated above that sometimes applies. If the leading digit in an uncertainty is small (a 1 or perhaps a 2), then it may be appropriate to retain one extra digit in the final answer. For example, an answer such as 17.6 ±1 is quite acceptable since one could argue that to round it to 18 ± 1 would be a waste of information.

Discrepancy ≡ *difference between two measured values of the same quantity.* A discrepancy may or may not be significant.

Uncertainty in a Difference: If the quantities x and y are measured with uncertainties δx and δy, and if the measured values of x and y are used to calculate the difference $q = x - y$, then the uncertainty in q is the sum of the uncertainties in x and y:

$$\delta q <= \delta x + \delta y.$$

Fractional Uncertainty: Another useful way of expressing uncertainty is to do so using what is called "fractional uncertainty." Fractional uncertainty is simply the actual uncertainty of a parameter divided by the best estimate magnitude of the same parameter. Thus,

$$\textit{fractional uncertainty} = \frac{\delta x}{|x_{best}|} \qquad (B.2)$$

Appendix B

Fractional uncertainty is used to tell us about the relative or percentage accuracy of a measurement. Note that, since both δx and x_{best} have the same units, fractional uncertainty is a unitless parameter. For example, if the length of a book is given as 234 ±1 mm, its fractional uncertainty is 1/234 = 0.004 (i.e., 0.4%). If the diameter of a pencil is given as 8 ±1 mm, its fractional uncertainty is 1/8 = 0.125 (12.5%). While both measurements have the same uncertainties, the fractional uncertainties differ by a factor of 35. The diameter of the pencil is much more uncertain than the length of the book.

Many parameters are determined by combining measurements. The perimeter of a building is the sum of all of the lengths of the edges of the building. The volume of a box is the product of the length, width, and height. The energy of a moving object = $1/2\, m\, v^2$ where m is the mass of the object and v is its velocity. When we want to know the uncertainty of a parameter, we need to know how to combine the uncertainties of the individual components to form the final uncertainty in the answer. Taylor has a number of rules which cover most possible combinations of calculations.

<u>Uncertainty in Sums and Differences</u>: If several quantities $x,...,\ w$ are measured with uncertainties $\delta x,...,$ δw, and the measured values are used to compute

$$q = x + ... + z - (u + ... + w) \qquad (B.3)$$

then the uncertainty in the computed value of q is the sum

$$\delta q \le \delta x + ... + \delta z + \delta u + ... + \delta w, \qquad (B.4)$$

of all of the original uncertainties. In other words, when one adds or subtracts any number of quantities, the uncertainties in those quantities always add.

Uncertainty in Products and Quotients: If several quantities $x,...,$ w are measured with small uncertainties $\delta x,...,$ δw, and the measured values are used to compute

$$q = \frac{x \cdot ... \cdot z}{u \cdot ... \cdot w} \qquad (B.5)$$

[where "$\cdot$" means "product" or "times"], then the fractional uncertainty in the computed value of q is the sum of the fractional uncertainties in $x,...,$ w given as follows:

$$\frac{\delta q}{|q|} \cong \frac{\delta x}{|x|} + ... + \frac{\delta z}{|z|} + \frac{\delta u}{|u|} + ... + \frac{\delta w}{|w|}. \qquad (B.6)$$

Briefly, when one multiplies or divides quantities, the **fractional uncertainties add.**

Note the use of $|\ |$ (absolute value) lines in the above expression. These are used to make sure that the uncertainties really do add. If parameter z were negative, for example, its fractional uncertainty would be subtracted from the total, reducing the uncertainty estimate of q. By using $|z|$, the fractional uncertainty in z would be added with the others, giving the true fractional uncertainty.

Measured Quantity Times Exact Number: If the quantity x is measured with uncertainty δx and is used to compute the product

$$q = Bx, \qquad (B.7)$$

where B has no uncertainty, then the uncertainty in q is just $|B|$ times that in x,

$$\delta q = |B| \, \delta x. \qquad (B.8)$$

Uncertainty in a Power: If the quantity x is measured with uncertainty δx and the measured value used to compute the power

$$q = x^n, \qquad (B.9)$$

then the fractional uncertainty in q is $|n|$ times that in x,

$$\frac{\delta q}{|q|} = \frac{|n|\delta x}{|x|} \qquad (B.10)$$

Often when measurements are combined, some of the parameters will be larger than their measured values while others will be smaller. If we add the uncertainties of all of the measurements together, we will overestimate the uncertainty in the final quantity. If the measured values are independent of each other, it is likely that some of the errors will cancel each other. Thus, we need to account for this in stating our uncertainties. Taylor continues with the following rules.

Uncertainties in Sums and Differences: Suppose that x, ..., w are measured with uncertainties δx, ..., δw, and the measured values are used to compute

$$q = x + \ldots + z - (u + \ldots + w). \qquad (B.11)$$

If the uncertainties in x, ..., w are known to be **independent and random**, then the uncertainty in q is the quadratic sum

$$\delta q = \sqrt{\delta x^2 + ... + \delta z^2 + \delta u^2 + ... + \delta w^2} \qquad (B.12)$$

of the original uncertainties. In any case, δq is never larger than the sum

$$\delta q \cong \delta x + ... + \delta z + \delta u + ... + \delta w. \qquad (B.13)$$

Uncertainties in Products and Quotients: Suppose that x, ..., w are measured with uncertainties δx, ..., δw, and the measured values are used to compute

$$q = \frac{x \cdot ... \cdot z}{u \cdot ... \cdot w}. \qquad (B.14)$$

If the uncertainties in x, ..., w are **independent and random**, then the fractional uncertainty in q is the sum in quadrature of the original fractional uncertainties.

$$\frac{\delta q}{|q|} <= \sqrt{\left(\frac{\delta x}{x}\right)^2 + ... + \left(\frac{\delta z}{z}\right)^2 + \left(\frac{\delta u}{u}\right)^2 + ... + \left(\frac{\delta w}{w}\right)^2}$$
$$(B.15)$$

In any case, it is never larger than their ordinary sum

$$\frac{\delta q}{|q|} <= \frac{\delta x}{|x|} + ... + \frac{\delta z}{|z|} + \frac{\delta u}{|u|} + ... + \frac{\delta w}{|w|}. \qquad (B.16)$$

Uncertainty in Any Function of One Variable: If x is measured with uncertainty δx and is used to calculate the function $q(x)$, then the uncertainty δq is

Appendix B

$$\delta q = \left| \frac{dq}{dx} \right| \delta x \qquad (B.17)$$

where dq/dx is the derivative of q with respect to x.

Uncertainty in a Function of Several Variables: Suppose that x, ..., z are measured values used to compute the function $q(x,..., z)$. If the uncertainties in x,..., z are independent and random, then the uncertainty in q is

$$\delta q = \sqrt{\left(\left(\frac{\partial q}{\partial x} \right) \delta x \right)^2 + ... + \left(\left(\frac{\partial q}{\partial z} \right) \delta z \right)^2}. \qquad (B.18)$$

In any case, it is never larger than the ordinary sum

$$\delta q \leq \left| \frac{\partial q}{\partial x} \right| \delta x + ... + \left| \frac{\partial q}{\partial z} \right| \delta z. \qquad (B.19)$$

In this notation, "∂" is the partial derivative of a parameter.

The above summarizes much of Taylor's information on error analysis and the measurement and statement of uncertainties. Much of it can be applied directly to radar meteorology. There is one problem, however, which Taylor did not explicitly address that is important for meteorological measurements. And that is the uncertainty of measurements on a logarithmic scale.

Uncertainties for Logarithmic Parameters: Many measurements that have to do with radar meteorology are made using logarithmic units. Among the parameters which are measured in logarithmic units are the following: transmitter peak and average powers;

receiver power; antenna gain; attenuation in the waveguide, radome and other system components; attenuation in the atmosphere; radar reflectivity factor Z; backscattering cross-sectional area σ. On the other hand, a number of other parameters related to radar are measured on a linear scale, including: wavelength; transmitter frequency; pulse duration; pulse repetition rate; distance from the radar. Unfortunately, in order to calculate some parameters such as radar reflectivity factor Z, we must combine both linear and logarithmic parameters together; the final answer is often expressed logarithmically.

[Note: At this point let me point out a shortcoming of the unimaginative radar meteorological community: The same symbol Z is usually used for both linear and logarithmic units of radar reflectivity factor. In this discourse, I have differentiated between linear and logarithmic units by using lower case letters for linear parameters and capital letters for logarithmic parameters. With printed text, it may be difficult to distinguish between these, unfortunately, so be careful and be observant. My experience in the use of "dBZ" has usually been to have the "Z" as a lower case letter. This would better follow the convention introduced herein, namely to use lower case z to represent linear reflectivities. The American Meteorological Society has currently settled on using the capital Z in dBZ, however, so I will follow their current system for "dBZ". It is certainly better than what they tried to foist onto the meteorological community a few years ago (i.e., "dB(Z_e)"!]

One of the primary definitions of logarithmic units is that of the decibel. If we have two "linear" powers p_1 and p_2 and take their ratio, we get the power ratio p_1/p_2 which is a unitless quantity. On a logarithmic scale, the logarithmic power ratio P is given by

$$P = 10 \, log_{10}\left(\frac{p_1}{p_2}\right), \qquad (B.20)$$

where the logarithm is to the base 10. Again, p_1/p_2 is unitless (or must be made unitless before taking the logarithm), but the logarithmic power ratio P defined above is expressed in "decibels" or dB.

Notice that when the ratio of two powers is taken, it becomes a relative measure rather than an absolute measure. That is, all we know is that one power is larger or smaller than the other and the magnitude of the difference between them. For example, if $p_1 = 100$ W and $p_2 = 50$ W, when we take the ratio of these we get 2. Given only the ratio, all we can say is that p_1 is twice p_2. We cannot say how big p_1 is unless we know p_2, and vice versa.

To make absolute power measurements using logarithmic units, we can set $p_2 = 1$ mW. Then the first power p_1 is compared to a standardized reference power (1 mW) and the logarithmic power P is then an absolute power defined by

$$P = 10 \, log_{10}\left(\frac{p_1}{1 \, mW}\right) \qquad (B.21)$$

where logarithmic power P is expressed in dBm; linear power p_1 must be in mW or converted into mW. Also, we can drop the subscript "1" because we are now dealing with only one power.

Anytime we see a parameter expressed in "dB", we should know it is a relative measure of something.

Whenever we see "dBm", we should know it is an absolute power.

When working on a linear scale, radar reflectivity factor z is expressed in units of (mm^6/m^3) where these units result from the definition of

$$z = \sum_{i=1}^{n} N_i D_i^6 \qquad (B.22)$$

and N_i is the number of drops of diameter D_i per unit volume (i.e., number/m^3); D_i is measured in mm. The logarithmic definition of reflectivity is

$$Z = 10 \log_{10}\left(\frac{z}{1\ mm^6 / m^3}\right) \qquad (B.23)$$

and the units of Z are dBZ. This is an absolute measure of reflectivity because it is relative to a reference value of 1 mm^6/m^3. Because it becomes cumbersome to include the term (1 mm^6/m^3) in the denominator each time we write reflectivity logarithmically, you may occasionally find the expression $Z = 10 \log_{10}(z)$ or even $Z = 10 \log z$ (omitting the base of the logarithms). As long as z is measured in mm^6/m^3, the answer will be correct. The term in the denominator is required mathematically, however, because you can take the logarithm of a number but the logarithm of a unit is meaningless (e.g., what is the meaning of the logarithm of a "meter"? Answer: it has no meaning!). Finally, the units used to express reflectivity logarithmically are dBZ, where the "Z" tells us that it is in decibels relative to a reflectivity of 1 mm^6/m^3.

Now, returning to the problem of expressing uncertainties, we can approach this in two different but related ways. One simple way is to write the

Appendix B

logarithmic value of reflectivity and its uncertainty in terms of the linear value and its uncertainty. We may write our expression as follows:

$$Z + \delta Z = 10 \log (z + \delta z).$$

Rearranging and solving for δZ gives

$$\delta Z = 10 \log \left(1 + \frac{\delta z}{z}\right) \qquad (B.24)$$

This last term on the right is the fractional uncertainty of the linear value of z. Thus, the uncertainty of the logarithmic value of Z is one plus the fractional uncertainty of the linear z but expressed in logarithmic units. In this case the answer is expressed in dB, not dBZ because it is a relative value rather than an absolute value.

We can take the same initial expression and solve for the fractional uncertainty of the linear reflectivity z in terms of the uncertainty in the logarithmic reflectivity δZ. Doing this gives

$$\frac{\delta z}{z} = 10^{\left(\frac{\delta z}{10}\right)} - 1. \qquad (B.25)$$

In this expression the first term to the right of the equals sign simply converts a logarithmic parameter into a linear one.

The second approach to expressing the uncertainty in logarithmic units can be obtained by following the expression given by Taylor (Eq. B.17) which uses the derivative of q with respect to x:

$$\delta q \ = \ \left| \frac{dq}{dx} \right| \ \delta x$$

where the absolute signs are again included for completeness. If we recognize that we must use the absolute value of this parameter, we can drop these and simplify our notation somewhat.

If we have

$$Z = 10 \ log_{10} \ z$$

where

$$z = z_{best} \pm \delta z,$$

then

$$q = Z \ = \ 10 \ log_{10} \ z$$

and

$$x = z$$

so

$$\delta Z \ = \ \frac{d(10 \ log \ z)}{dz} \ \delta x$$

From math tables,

$$d(log_a u) \ = \ u^{-1} \ log_a e \ du$$

where $a = 10$ and $u = z$, so

$$d(10 \ log_{10} \ z) \ = \ 10 \ d(log_{10} z)$$

$$= \frac{10}{z} \log_{10} e \, dz \, .$$

Then

$$\delta Z = \frac{\dfrac{10}{z} \log_{10} e \, \delta z}{\delta z} \, \delta z$$

giving the final expression for logarithmic uncertainty (in decibels) in terms of fractional uncertainty

$$\delta Z = 4.343 \frac{\delta z}{z} \, .$$

Recalling that we need to calculate uncertainty to only one significant digit, we can simplify this expression into our working version for converting fractional uncertainties into logarithmic uncertainties. This is

$$\delta Z \cong 4 \frac{\delta z}{z} \qquad (B.26)$$

Finally, we can calculate both the value of Z and its uncertainty and express it as follows:

$$Z = Z_{best} \pm \delta Z,$$

where

$$Z_{best} = 10 \log_{10} z_{best}.$$

Thus, the expression used to calculate logarithmic reflectivity and its uncertainty from the corresponding linear values is given by

$$Z \pm dZ = 10 \log z_{best} \pm 4 \frac{\delta z}{z_{best}}.$$

We can perform a similar set of manipulations to derive an expression for the uncertainty of linear z based on Z and its logarithmic uncertainty. The result of this manipulation is the following expression:

$$\frac{\delta z}{|z|} = 0.2303 \, \delta Z$$

which can again be approximated to one digit of accuracy with

$$\frac{\delta z}{|z|} \cong \frac{\delta Z}{4} \qquad (B.27)$$

The two different expressions for δZ (Eqs. B.24 and B.26) and the two for $\delta z/z$ (Eqs. B.25 and B.27) give virtually the same answers as long as the uncertainties in the complementary form (logarithmic or linear, as the case may be) are not excessively large. For example, as long as the fractional uncertainty $\delta z/z$ is less than about $\pm 50\%$, the logarithmic uncertainty δZ will be within about 0.4 dB of the same value. Consequently, it really doesn't matter which expression is used for most conversions. If extremely large uncertainties are involved, then the second versions might be best (i.e., those based on Taylor's approach). Certainly, the easiest relationships between fractional uncertainties and logarithmic uncertainty are Eqs. B.26 and B.27. Knowing either of these makes it possible to convert very quickly and with adequate precision for our purposes. They are so simple, you should be able to

Appendix B

multiply or divide by four in your head to convert from linear to logarithmic uncertainties most of the time.

Example for Point-Target Radar Equation: Let us apply the above information to a radar example. Suppose the UND C-band radar (see Appendix D for complete specifications) detects a point target at a certain distance, what is the uncertainty of the backscattering cross-sectional area of the measured target? For this calculation we need to use the point target form of the radar equation given as

$$p_r = \frac{p_t\, g^2\, \lambda^2\, \sigma}{64\, \pi^3\, r^4}$$

Parameter	Value (units)	Uncertainty (units)
P_r	-85 dBm	0.7 dB
p_t	250 kW	10%
G	43.75 dB	0.5 dB
λ	5.4 cm	0.01 cm
r	10 km	0.1 km

where p_t is transmitted power, g is antenna gain, λ is wavelength, σ is the backscattering cross-sectional area of the target, and r is range to the target. Since we know or will measure p_r, let us solve the equation for σ, giving

$$\sigma = \frac{64\, \pi^3\, r^4\, p_r}{p_t\, g^2\, \lambda^2}$$

Let us assume the following values and uncertainties. In this table the values and uncertainties are given in the units most commonly used for their measurement.

Notice that two of the values are given in logarithmic and three in linear units. Further, p_t has a percentage uncertainty associated with it, not an absolute uncertainty. The first step is to convert all parameters and uncertainties into either linear or logarithmic units. Since linear units are most frequently used, let us convert to linear values. This gives

Parameter	Value (units)	δq	$\dfrac{\delta q}{q}$
p_r	$10^{-8.5}$ mW	$5.53 \cdot 10^{-10}$ mW	0.17
p_t	250 kW	25 kW	0.10
g	$10^{4.375}$	2894	0.12
λ	5.4 cm	0.01 cm	0.002
r	10 km	0.1 km	0.01

First of all, we can see that the uncertainty in wavelength is far smaller than any of the others, and we can safely ignore it. We also see that the uncertainty in received power (17%) is the largest, followed by the uncertainties in gain (12%), transmitted power (10%), and range (1%).

Let's combine these to calculate the fractional uncertainty in backscattering cross-sectional area σ. The equation used to do this is a combination of a couple of the equations given earlier. It is:

$$\frac{\delta \sigma}{|\sigma|} = \left(\left(\frac{\delta p_r}{p_r} \right)^2 + \left(\frac{\delta p_t}{p_t} \right)^2 + 2\left(\frac{\delta g}{g} \right)^2 + 4\left(\frac{\delta r}{r} \right)^2 \right)^{0.5}$$

$$= [0.0289 + 0.010 + 0.0288 + 0.0004]^{0.5}$$

$$= 0.26$$

$$= 26\%.$$

Obviously, we could also have ignored the uncertainty in our range measurement in this calculation. In fact, one of the advantages of doing an error analysis is that it will point out where the greatest and least errors are in your measurements. From this it is possible to direct your efforts to improving those parameters which are least well known. Another point to make about the above equation is that gain and range terms have factors of 2 and 4, respectively, in front of them. This is required because gain is squared and range is raised to the 4th power in the equation to calculate σ.

This is really all we need to know about the uncertainty of our measurement. We can, however, apply it to the actual value to get an absolute uncertainty. First we need to calculate σ. Plugging values into the equation, we find that

$$\sigma = \frac{64\,\pi^3\,r^4\,p_r}{p_t\,g^2\,\lambda^2}$$

$$= \frac{64\,\pi^3\,(10\ km)^4\,(10^{-8.5}\ mW)}{250\ kW\ 23714^2\,(5.4\ cm)^2}$$

$$= \left(\frac{1.302\ 10^{-11}\ km^4\ mW}{kW\ cm^2}\right)\left(\frac{1\ kW}{10^6\ mW}\right)\left(\frac{(10^5\ cm)^4}{km^4}\right)$$

$$= 1302 \text{ cm}^2.$$

Now we can combine this with our uncertainty to get

$$\sigma = 1302 \text{ cm}^2 \pm 26\%$$

$$= 1302 \pm 338 \text{ cm}^2$$

$$= 1300 \pm 300 \text{ cm}^2$$

This, then, is the final answer.

Just for completeness, let us convert this answer to logarithmic units to see what it looks like.

$$\Sigma = 31.1 \text{ dB}\sigma \pm 1.0 \text{ dB}$$

$$= 31 \pm 1 \text{ dB}\sigma.$$

In this expression I have used capital sigma (Σ) for the logarithmic symbol for backscattering cross-sectional area.

ALL I SAID WAS: "WHAT'S A FEW DB AMONG FRIENDS."

Tools and Toys

"A workman is only as good as his tools" an old saying goes. Modern radar is certainly a good tool. But even with the most modern radar available, it often seems that there are a few small things that make using it even easier to do certain jobs. [Alternatively, tools have a way of becoming obsolete as new and better tools become available.]

There are a couple tools (toys?) I developed while working with a small X-band radar in Kericho, Kenya, a number of years ago that I found indispensable. One was a circular slide rule that was used to convert received powers and distances to storms into radar reflectivity factors. The second was a speed calculator for converting distances traveled over known time intervals into speeds. Perhaps the time is past when these kinds of mechanical "computers" are useful. Nevertheless, since some radar meteorologists still have to manually determine speeds and forecast arrival times of storms over a particular location and may still have to convert analog signal strength values into reflectivities, there may still be a use for these. In the case of reflectivity, in particular, there may be little need for a mechanical way to make calculations. Modern color displays coupled with computer processors show

Appendix C

reflectivities perfectly well. Older radars with monochrome displays, however, usually cannot resolve reflectivities quantitatively. It was with such a radar that the circular slide rule was extremely useful.

Reflectivity Slide Rule:

Figure C.1 shows the components for a circular slide rule which mechanically solves the radar equation of the form

$$z = c \, p_r \, r^2$$

where p_r is the received power from a target, r is the range to the target, c is the radar constant, and z is radar reflectivity factor in mm^6/m^3. In this version of the radar equation, the constant c is given by

$$c = \frac{1024 \ln(2) \lambda^2}{\pi^3 \, p_t \, g^2 \, \theta \, \phi \, h \, |K|^2}$$

where λ is wavelength, p_t is the peak transmitter power (mW), g is antenna gain, θ and ϕ are horizontal and vertical beamwidths of the radar antenna (in radians), h is radar pulse length, $|K|^2$ is the dielectric constant term (usually taken as 0.93). Also, the units must be such that all units cancel and z is in mm^6/m^3. This constant is slightly different for each radar. The value of c used on the slide rule is $10 \log(c)$. One key to using this slide rule is placing a mark on the "radar constant" scale at the bottom of the inner disk at the correct location. Several radars are already marked on the scale. Notice that they are all clustered near one location on the scale. While this may at first seem surprising, it is not unreasonable that numerous radars designed for detecting meteorological targets would be fairly similar.

To construct the slide rule, simply cut out the two circles and place some kind of pin or rivet or other central axis in the middle of both wheels. It may be desirable to glue the circles onto some kind of stiff card stock or other more rigid material to add strength to the device. That would also give more support for the center pivot point.

Another key to using the slide rule is in knowing the receiver power. Older radars and some contemporary radars have manual gain controls on them to adjust the sensitivity of the radar receiver. Gain controls operate much like a volume control on a radio or television. By turning the knob to the maximum sensitivity position (usually full clockwise), the receiver will amplify the received powers as much as possible and the radar will "see" the weakest echoes possible. By reducing the gain somewhat, weaker echoes will disappear but stronger echoes which may have saturated the receiver, will still be detectable. If the gain control is reduced further, even strong echoes will disappear. The "proper" position of the gain control knob is to reduce the gain until the echo being measured is just barely detectable (i.e., it is a "minimum detectable signal"). When an echo is just barely detectable, the gain control setting is then related to received power.

For this technique to be quantitatively useful, we must calibrate the receiver gain control and mark or note the position corresponding to various power levels. Using a signal generator, we can inject a signal of known strength into the radar's directional coupler. Then we adjust the receiver gain control until the signal is just barely detectable (MDS). Mark or record that gain control position with the corresponding power. Repeat this procedure using as wide a variety of powers as needed to completely cover the range of values the radar is capable of receiving. The received power must account for all of the losses in the system as described in Chapter 12, Quantification.

Some radar displays have more than one way to adjust the display. For example, some will have an "intensity" or "contrast" control as well as a gain control. The intensity/contrast control adjusts the range of brightness the display tube will cover. If you plan to make quantitative measurements from such a display, it is important that you always set the intensity control in a standardized way. One way to do this is to adjust the receiver gain control to the minimum position (full counterclockwise) and then adjust the intensity until the sweep of the radar is just barely detectable; then reduce the intensity control a slight amount more so the sweep is no longer visible [Note: This adjustment is best made with the antenna scanning; in any case, perform your calibration and make your measurements using the same conditions]. Once the background intensity has been adjusted, turn the receiver gain control up until it looks "good". This is also a matter of personal preference to some extent; I prefer to have the screen showing at least some background noise.

A modification to this calibration procedure that can simplify the use of the slide rule is to add gain control settings directly on the slide rule instead of using received power marks. For example, if the gain control is labeled with numbers from 0 to 10, you could put these numbers on the slide rule along with or in place of the received power scale on the top of the calculator. Then all you need to do to make a reflectivity calculation is to note the position of the gain control in its own number system for a minimum detectable signal, and use that number of the slide rule. That can save an intermediate step.

One way to get the correct receiver gain number to add to the slide rule is to plot a receiver calibration graph as follows. Plot received power as the abscissa of the graph and the receiver gain control as the ordinate. Then transfer the corresponding gain control setting for every 10-dB step in receiver power from MDS through saturation onto the slide rule. Many gain controls will not be very linear (at least on a logarithmic power scale), so you may find widely spaced numbers at one part of the slide rule and closely spaced numbers somewhere else.

Once the slide rule has been correctly labeled, reflectivity measurements can be made with it. Since I had performed calibrations with the antenna scanning, I also made the measurements with the antenna scanning. However, to reduce the time needed to make a reflectivity measurement for a specific storm, I would manually scan the antenna back and forth across the storm while simultaneously adjusting the gain control to get a minimum detectable signal. Once the strongest point of the storm had been identified and logged, I would usually then return the gain control to full sensitivity and scan the antenna up and down to measure the storm top.

Speed Calculator:

The component parts of the speed calculator are shown in Fig. C.2. The speed calculator mechanically solves the equation

$$v = \frac{d}{t}$$

where v is speed, d is distance, and t is time. Actually, given any two of these parameters allows us to calculate the third. Initially, we would measure travel distance and time and use the calculator to give speed. But once the speed of movement of a storm (or other object) is known, we can use the calculator to determine either the time it will take to travel a known distance or, given a time, we can determine how far the object will travel.

This device is made of two pieces connected at a pivot point. The base piece contains both a distance scale near the outer edge and a family of speed lines across the major part of the face of the calculator. The main body piece also has a pointer attached to it. The pivoting piece and its pointer also has a time line. The distance from the pivot point to the end of the points is adjusted to match the scale of the radar display from which measurements will be made. In fact, it is possible to construct such a speed calculator with adjustable length pointers so that the same calculator can be used with different scale displays (Rinehart, 1980).

To use the calculator, mark the location of an echo of interest on the display using a grease pencil or a felt-tipped pen; also record the time. Then, some time later, mark the new location and record its time. Measure the distance traveled by adjusting the angle between the pointers on the speed calculator until they match the distance traveled. Determine the travel time. Next to the travel time on the moveable pointer piece, read the travel speed on the speed scales under this point.

Once a speed for a target is known, it becomes possible to forecast the time of arrival of the target at some future location. To make a forecast, span the distance from the present location of the target and the distance to the point of interest. Then, find the speed line on the body that corresponds to the target's speed.

Read the time it will take to arrive at the new location where the time line intersects the speed line. These kinds of forecasts assume the target will continue to move at the same speed it had up until the present time. Storms are generally reasonably forecastable, but they do change. New echoes have a way of forming where not anticipated. Old echoes have a way of dying unexpectedly. Both happenstances can make reasonable forecasts completely wrong. So, don't attempt extremely long forecasts, and revise your forecasts as new information becomes available.

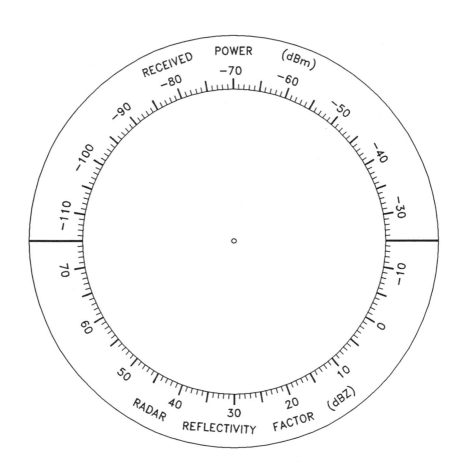

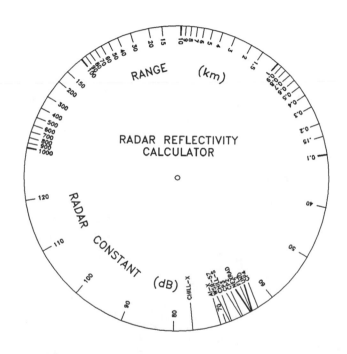

Appendix C

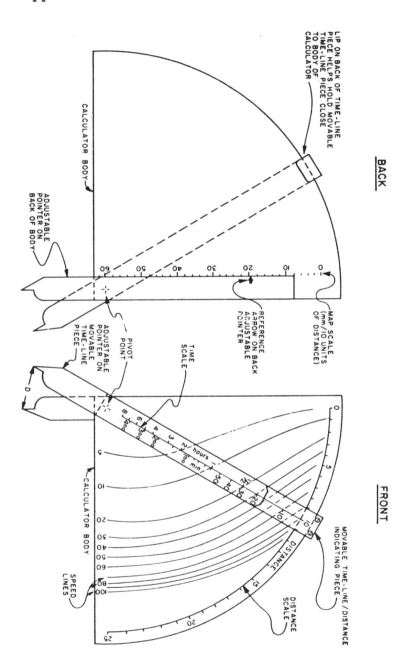

348

R. E. Rinehart and Barry Ross Rinehart, 1997

Appendix D

Appendix D: Specifications and Zmin

	Radar:	UND	FL2	WSR-88D	WSR-57
Antenna	(units)				
Diameter	ft	12	28	28	12
	m	3.66	8.53	8.53	3.7
Beamwidth	°	0.99	0.96	1	1.8
Gain	dB	43.75	46.1	45	38.5
Polarization		H	H	H	H
Rotation rate					
Max	°/s	24	30	36	
Typical	°/s	5 to 20	25		
Transmitter					
Band		C	S(C)*	S	S
Frequency	GHz	5.55	2.865	2.8-3.0	3
Wavelength	cm	5.4	10.46	10.71	9.99
Peak power	kW	250	1000	1000	500
Pulse length	μs	0.6, 2	0.65	1.57, 4.5	0.25, 4
PRF	Hz	250-1200	700-1200	318-1403, 318-452	200-800
Receiver					
Bandwidth	MHz	0.3	1.3	0.63	0.75
MDS	dBm	-106.5	-107	-113	-110
Noise figure	dB	4	1.5		
A/D converter	bits	10	12		
Clutter suppress.	dB	30, 40	50	30 to 50	
No. range gates		1024	400, 800		
Gate spacing	m	25, 50, etc.	120, 240, etc.	250	
Radar constant***	**dB**	**64.5**	**64.7**	**58.4**	**69.2**
Zmin @ 50 km	dBZ	-8.0	-8.3	-20.7	-6.8

*** using longest pulse duration   * S, '85-'90; C, '91
for use in $Z = P_r + 20 \log(r) +$ Radar Constant

for selected meteorological radars

CP2	CP3/4	CHILL	NSSL	Atmos. Inc.	TDWR	ASR-9
28	13.8	27.89	30	3	25	(»16 by 9)
8.53	4.2	8.5	9.14	0.91	7.6	(»4.9 by 2.7)
0.96	1	0.96	0.9	4	0.5	1.4 by 5
42.5	43	43.3	46	34	51	34
H, V	H	H, V	H, V	H	H	V or C
14	25	30	18	60	30	72
10			6 to 12	60		72
S,X**	C	S,X**	S	X	C	S
2.809	5.5	2.73	2.735	9.375	5.6-5.65	2.7-2.9
10.67	5.45	10.98	11	3.21	5.3	10.7
1200	1000	1000	500	40	250	1100
0.2-1.5	1	.25, 0.5, 1, 1.7	1	2.35	1.1	1
400-1667	750-1667		814, 930, 1085, 1302	400	235-2000	8 @ 950 & 10 @ 1200
1.4	1.4	0.3	1		1	
-109	-110	-110	-107	-103.9	-106	-114
3	3	4		10?	1.4	4.1
		8	12	(analog)	12	12
No	Yes		20 to 50			60
256-1024	<1500	1024-4096	756			960
50-2500	75-2500	27.5, 75, 150	150, 180, 210, 240			150
67.7	**63.0**	**66.4**	**67.1**	**74.7**	**58.4**	**58.0**
-7.4	-13.0	-9.6	-6.0	4.7	-13.6	-22.0

** dual-wavelength radars; specs given for S-band portion

Appendix D

The radars listed on the previous pages are operated by the following organizations:

UND: Department of Atmospheric Sciences, University of North Dakota, Grand Forks, North Dakota. Radar is a modified WSR-74C built by Enterprise Electronics, Corp.

FL2: Massachusetts Institute of Technology Lincoln Laboratory, Lexington, Massachusetts. Radar built using transmitter from ASR-8 and other components built by Lincoln Laboratory.

WSR-88D: National Weather Service, National Oceanic and Atmospheric Administration, Department of Commerce; United States Air Force, Department of Defense; Federal Aviation Administration, Department of Transportation; all headquartered in Washington, D.C. Radar built by Unisys Corp.

WSR-57: National Weather Service, National Oceanic and Atmospheric Administration, Department of Commerce, Washington, D.C.

CP2, -3, and -4: Atmospheric Technology Division, National Center for Atmospheric Research, Boulder, Colorado.

CHILL: University of Chicago/Illinois State Water Survey, Chicago and Urbana, Illinois, respectively, until 1990; Department of Atmospheric Sciences, Colorado State University, Fort Collins, Colorado, after 1990.

NSSL (Cimarron): National Severe Storms Laboratory, National Oceanographic and Atmospheric

Administration, Department of Commerce, Norman, Oklahoma. Radar is a modified FPS-18 radar.

Atmos. Inc.: Atmospherics Incorporated, Fresno, California. Radar primarily a ship-board radar with modifications, especially for the antenna.

TDWR (Terminal Doppler Weather Radar) and ASR-9: Federal Aviation Administration, Department of Transportation, Washington, D.C.

Appendix E: Characteristics of two air-traffic control radars (FAA, 1991 for ARSR-4 and Skolnik, 1980: *Radar Handbook*, with permission of McGraw-Hill, Inc., for ARS-8)

Parameter	Unit	ARSR-4	ARS-8
Frequency band		L	S
Frequency	MHz	1215-1400	2700-2900
Instrumented range	km	460	111
	n mi	250	60
Peak power		62.5 kW per channel; 3 channels	1.4 MW
Average power	W	2592	875
Noise figure	dB	4	4
Pulse width	μs	150;	0.6
Compressed duration:		1.4	
Pulse repetition frequency	Hz	288 average	700-1200 1040 ave.
Antenna rotation rate	RPM	5	12.8
Antenna size	m	5.12 by 3.66	4.9 by 2.7
Azimuth beamwidth	°	1.4	1.35
Elevation coverage	°	30	30
Antenna gain	dB	37.7 low 35.5 high 40 ave. receive	33
Polarization		linear or circular	vertical or circular
Blind speed	kt		800
MTI improvement factor	dB		34

Appendix F

Filter-Paper Technique

The purpose of this Appendix is to describe how to measure drop-size distributions of rain using the filter paper technique. However, in order to do this, we must first calibrate the system so that meaningful measurements can be made.

Filter paper dusted with methylene blue dye and exposed to water will turn dark blue. By dropping water drops of known sizes onto dye-treated filter paper, we can determine the relationship between the spot sizes on the paper and the diameters of the incident water drops. The size of the blue spot on the paper is related to the volume of water in a given drop. Thus, as the drop diameter increases, hence increasing the drop volume, the spot diameter on the filter paper will also increase. There are two ways we can calibrate this kind of measurement system: empirically and theoretically.

Empirical calibration: In the empirical approach we simply drop water drops of known diameter onto the filter paper and determine an empirical relationship between drop diameter and spot diameter. This relationship, either in the form of an equation or as a carefully plotted graph, can then be used to determine drop diameters from spot diameters. It's really that simple. The problems arise, of course, in knowing the size of the drops impinging upon the paper. This will be covered in more detail in the Procedure section.

Theoretical calibration: Assume a water drop of diameter D_d that falls and hits a piece of filter paper. The volume of the drop is given by

$$V_d = \frac{4\pi}{3}r_d^3 = \frac{\pi}{6}D_d^3$$

where subscript d refers to the drop, r is the radius of the drop, and D is the diameter of the drop.

If the paper is at all porous (and all filter paper must be porous or it would not be used), then we can associate two thickness to the paper (see Fig. F1). The first thickness is the simple one which we get when we measure its thickness; let us call this t. The second thickness we can call the compressed thickness k; this is the thickness the paper would have if we could squeeze all the air out of the paper, producing a solid mass of paper. Obviously, the compressed thickness must be smaller than the total thickness.

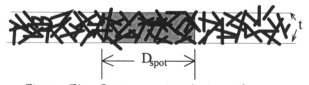

Figure F1. Cross-sectional view of a piece of filter paper showing the individual fibers of the paper, its overall thickness t and the volume filled by a water droplet with spot diameter D_s.

When a drop of water hits a piece of filter paper, the water will spread out until it fills the pores of the paper to the same volume as the drop. This assumes that there are no air pockets within the volume occupied by the water. If this is correct, then the volume of the drop must equal the volume of paper formerly occupied by air (see Fig. 1). If the total thickness is t, and the compressed thickness is k, then the thickness which was air is $(t-k)$. If the drop forms a circular spot on the paper, it will occupy a cylindrical volume given by

$$V_s = \text{spot area} \cdot \text{air thickness}$$

$$= \pi r_s^2 (t - k) = \frac{\pi}{4} D_s^2 (t - k) \qquad \text{(F.1)}$$

where the subscript s refers to the spot.

If we equate the volume of the drop and the volume of the spot, we can solve for the drop diameter D_d, giving

$$D_d = \left(\frac{3}{2} (t - k) D_s^2 \right)^{1/3} \qquad \text{(F.2)}$$

$$= \left(\frac{3}{2} (t - k) \right)^{1/3} D_s^{2/3} \qquad \text{(F.3)}$$

This is our theoretical calibration. The parameters we need to know on the right side are drop diameter D_s, paper thickness t, and compressed thickness k. We can directly measure both D_s and t. How can we get k?

There are probably a number of ways to get k. One might be to beat the paper as thin as possible and measure k directly. This would assume that we succeeded in getting all the air out and that the resultant paper was uniformly thick. Another would be to weigh the paper, fill it with water and weigh it again, and somehow use the volume of water in the paper to get k. Another is to drop a number of drops of known size onto the paper, measure D_s corresponding to the known D_d and solve for k. This is perhaps the easiest way to get it. In fact, you could calculate k for every D_s-D_d pair of data you collect, since you know both of these for all trials. This would allow an estimation of the uncertainty of k, and knowing the uncertainty of a parameter can be useful.

No matter which calibration is used, the result is that we should be able to use filter papers to measure raindrops. Then we can use the resulting drop-size distribution to calculate a number of parameters. Raindrops falling from real rainstorms come in a variety of sizes. Typically, there are very many small raindrops and progressively fewer larger ones. Knowledge of the drop-size distribution makes it possible to relate the measured drop diameters to other parameters such as the radar reflectivity factor Z, rainfall rate R, and liquid water content M of the storm.

Given a size distribution, we can calculate radar reflectivity factor using the familiar

$$z = \frac{\sum_i^n N_i D_i^6}{V_i} \qquad (F.4)$$

where N_i is the number of drops of diameter D_i, V_i is the volume sampled (which is a function of the drop terminal velocity), and the summation is carried out over all n drops in the sample volume.

In order to use our drop-size distribution, we need to know the volume sampled for each diameter size interval. This is simply the product of the distance a raindrop falls during the measurement times the area over which the collection is made. We will assume that the raindrops we sampled are neither in an updraft nor in a downdraft but rather were falling at their terminal velocity.

For terminal velocity, let us use the relationship given by Beard (1985) given as follows:

$$V = V_o \left(\frac{\rho_o}{\rho} \right)^m \qquad \text{(F.5)}$$

where V is the terminal velocity of a raindrop of diameter D (measured in mm), ρ is the air density at the location where the drops are falling,

$$\rho_0 = 1.194 \ kg/m^3,$$

$$m = 0.375 + 0.025 \ D, \qquad \text{(F.6)}$$

and V_o is given by

$$V_o = exp(B_0 + B_1X + B_2X^2 + B_3X^3) \qquad \text{(F.7)}$$

and V_0 and V are in cm/s; $X = \ln(D)$, $B_0 = 5.984$, $B_1 = 0.8515$, $B_2 = -0.1554$, and $B_3 = -0.03274$. These empirical relationships fit the measurements of drop terminal velocity made by Gunn and Kinzer (1949) to about 1%.

Appendix F

The sample volume at each diameter interval will be equal to the area of the filter paper exposed to the rain times the distance each drop can fall during the exposure time. This distance is simply the terminal velocity given by the appropriate solution to the above equations times the exposure time. Thus, we can write our equation to calculate radar reflectivity factor as

$$z = \frac{\sum N_i D_i^6}{v_i A t} \qquad (F.8)$$

where v_i is the terminal velocity of a drop of diameter D_i, A is the area of the filter paper exposed to the rainfall, and t is the duration of the exposure.

For rain rate, we need to know how fast rain is accumulating at the surface. Rain rate is usually given in units of depth per unit time (e.g., mm/h). Alternatively, we can think of rain rate as the rate of accumulation of a given volume of water per unit area per unit time. The volume of water is given by the summation of the volumes of all the individual raindrops sampled. Thus, rainrate R is obtained by dividing the volume of water hitting the area by the area of the sample and the duration of the exposure, giving

$$R = \frac{\sum \frac{\pi}{6} N_i D_i^3}{A t} \qquad (F.9)$$

where each of the terms have the same meaning described before. Note that the terminal velocity of the raindrops does not affect rainrate calculations because the only thing that matters is those drops that actually hit the ground; i.e., we need to know the sample *area* but not the sample *volume*.

The liquid water content of a sample is the mass of water in a unit volume of space, typically given in grams of water per cubic meter of space. This is determined from the drop-size distribution by determining the volume of water contained by all raindrops in the volume sampled by each drop size. Thus

$$M = \frac{\rho_w \pi \sum N_i D_i^3}{6 v_i A t} \qquad \text{(F.10)}$$

where the terminal velocity v_i again appears and ρ is the density of water ($\rho = 1$ g/cm^3); including ρ in the equation converts the *volume* of water into the *mass* of water.

Apparatus:

To use the filter paper technique and to calibrate it, we need to use a number of different materials. The following list can be used as a guide, but be aware that you can substitute pretty freely for things that you find more conveniently available. In particular, any kind of filter paper can be used as long as you calibrate it correctly. The template given later, however, applies specifically to Watman No. 1 filter paper; it may give incorrect diameters if used with other kinds of filter paper, so be careful. Similarly, any water-sensitive dye can be used. So, here is a list of items:

Watman No. 1 filter paper, 18.5-cm diameter; methelyene blue dye (powered); newspaper; rubber gloves; plastic apron; dust masks; old clothes; soap, water and paper towels for clean up; clear plastic (acetate); micrometer; millimeter ruler; water; Ivory soap; eye droppers with small openings, at least two sizes

Procedure:

The following procedure may be modified somewhat from that given to suit your own particular situation. It should cover the procedure in enough detail, that you can make whatever modifications are needed to get the job done.

> *CAUTION:* This is a potentially messy procedure. Good lab technique is mandatory to avoid getting the methylene blue dye all over yourselves and the lab. Please exercise care in any steps that bring you into contact with the methylene blue dye. As the old saw goes: Cleanliness is next to Godliness; Rinehart's modification: Cleanliness is also closer to an A. Please be careful. Incidentally, there are no known health hazards involved in using methylene blue dye, but it certainly is an esthetic hazard (unless you like blue fingers, blue clothes, blue rooms, blue etc., etc., etc.). It would be wise to wear old cloths for this procedure to avoid ruining anything of value.

There are several steps involved in the calibration procedure. One is to determine the volume of a drop produced by a specific dropper. A second is to prepare the filter papers for use in the experiment. The third step is to expose the papers to drops of known sizes.

In order to use the theoretical approach to the calibration, we need to know the thickness of the paper.

Using a micrometer, measure the thickness of one or more single sheets of the paper (at one or more locations on each sheet) to determine both the thickness and variability of thickness. Alternatively, if a micrometer is not available, measure the thickness of many sheets and determine the thickness of the individual sheet by dividing by the number of sheets measured together. In either case, determine the uncertainty of your measurement.

To determine the size of drops produced by a given dropper, drop a known number of drops into a graduated cylinder. The drop volume can simply be determined by dividing the measured total volume by the number of drops used. Use enough drops in the graduated cylinder so that the total volume is on the order of 2 cm³. Read the actual volume to the nearest 0.05 cm³. That will give an uncertainty of about 2% which should be adequate.

Put a cotton ball over the mouth of the methylene blue bottle and get a small amount of powder on the ball. Then rub this evenly over one side of a piece of filter paper, being sure to cover the entire area with a light, uniform amount. It is not necessary to make the paper turn dark in color; a very light blue coating is sufficient. Make as many filter papers as you think you will need for your future measurements and for the calibration procedure; the calibration will likely use 5 to 10 papers.

There are at least two methods that can be used for empirically calibrating the filter papers. The first is to simply drop a number of drops from a given dropper onto a piece of filter paper (ten drops from each dropper should suffice). Do this for each different dropper size available. The second method is to use a single dropper that produces small drops, and drop multiple drops onto the same spot in quick succession (see Fig. F2). For example, make one spot with one drop, another spot using two drops, another spot using three, etc. Continue this until the spots are large enough so that they cover the largest possible drop of interest or until they become unusable (as from becoming irregular in shape). When dropping the drops, make sure there is enough space between drops so that they do not overlap or touch each other.

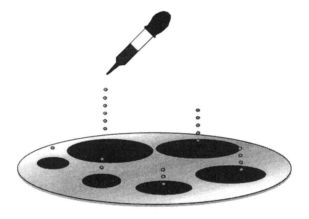

Figure F2. To increase the range of drop sizes available, drop more than one drop onto the same spot. The total volume of water dropped can be used to calculate the equivalent diameter of a single, larger drop that would have produced the same spot size.

One of the problems in this calibration is getting really small drops. One way to get slightly smaller drops than might otherwise be obtainable is to add a small amount of Ivory soap or some other kind of product to the water. This lowers the surface tension of the water and lets the drops release from the dropper sooner and with smaller diameters. Since the spot diameter on the filter paper is determined by the volume of water, the impurities added should not change the calibration significantly. To know what size the drops are, you will also have to use the same soapy water to count drops/measure volume using the graduated cylinder. Be sure you do not use one kind of water in the graduated cylinder and another (soapy or nonsoapy) on the filter paper; calibrate and drop using the same kind of water.

When exposing the dye-covered filter paper with water drops, make sure that the paper is supported by a non-water-absorbing material or hold it above the table while water is dropped onto it. This is necessary so that no water from the drop is lost to other materials, thereby reducing the available volume of water.

Once a filter paper has been exposed with water drops, place it aside somewhere to dry. Then place it under clear acetate or plastic to minimize the contact with the methylene blue dye while you make measurements of the spot sizes.

Measurements and Calculations:

Measure the spot sizes for each of your trials and determine both average size and variability. For the spots created by dropping multiple drops onto the same spot, you will have to calculate the diameter of the equivalent spherical drop which would have had the same volume as the total of n drops of know diameter.

Using the data collected in your calibration, you can either graph the results and use this for your calibration, fit a curve to the data points and use this, or determine the constant k and complete the theoretical calibration.

Appendix F

Exposing the filter paper: The procedure is conceptually quite simple. Using the filter papers you have already prepared, place a filter paper onto a rigid, nonabsorbing material such as a piece of plastic or other suitable material and tape, staple or clip it in place. Be careful not to cover too much of the paper with tape and/or clips; if too much area is covered during the exposure, you can account for this by estimating the actual exposure area available for raindrops to hit. The filter paper should be attached so it will remain in place during an exposure even if there is wind present (the usual case during rain). Thus, you should hold it in place at least three or four places around the edge.

When exposing the filter paper to the rain, there are two possible ways to do it. The first is to hold the filter paper perfectly level. The second is to tilt the paper so that it is aimed into the wind. Depending on how you need to use the results, either might be used. But exposing the papers horizontally gives the correct number of drops for each diameter size, so I would argue in favor of that method (Rinehart, 1982).

The exposure time for your filter paper will have to be a matter of personal judgement. If your exposure time is too short, you will have difficulty in getting a good measure of the actual exposure time (large fractional uncertainty). Also, you will not get a large enough sample of raindrops, especially for the larger sizes. On the other hand, if your exposure is too long, you will have so many drops on your filter paper that they will overlap. A small amount of overlap is fine (and maybe even desirable - it could suggest an optimum number of drops have been sampled). Further, the longer the exposure, the more drops there will be to count and measure. Typical exposures might be something like 30 s to 1 min for very light rain down to only a sec (or less if you can measure it well enough) for a heavy thunderstorm rainfall. You should think about how long you will expose each paper before you start so you can be prepared to stop the sample at the appropriate time.

To exposure your filter paper, get out into the rain with some kind of protective cover over the filter paper. Another piece of plastic or a piece of cardboard or something larger than the filter paper would do fine for this. When you are ready to make your exposure, remove the cover piece quickly so the entire filter paper gets as close to the same duration of exposure as possible. When your exposure time is over, cover the paper quickly. Record the exposure duration.

Measuring the raindrop diameters: When you get ready to measure the spot diameters, there are a number of ways to proceed. One is to measure every drop individually and generate a very large table of raindrop diameters. Actually, you will probably end up measuring drops only to a certain amount of precision, say 0.1 mm, 0.5 mm, or some other convenient interval. Then you can count or tabulate all drops that fall into that interval. Another way to get the diameters is to generate a clear plastic calibrated overlay. This takes a little while to produce, but it would speed up the measuring process enough to be a time saver if you have any reasonable number of drops to measure. If you used Watman No. 1 filter paper, the template provided in Fig. F4 can be copied onto transparency material and used for your measurements. Make sure your copy is exactly the same size as the original.

If your exposure time was too long and you ended up with many drops, you could subdivide the filter paper into halves or quarters or whatever is needed to get the number of drops to a more easily managed amount. Start by counting the largest drops. Continue working down toward smaller and smaller drop sizes. When the number of drops counted in a particular size interval exceeds about 50, divide the area in half and count all smaller drop sizes over one half of the initial area. If you reach the point where you again have about 50 of some size, divide the area into halves again and count the remaining drop sizes over one of these 1/4-sheet areas. The goal here is to insure that you count at least 25 drops of each diameter for the smaller size raindrops.

Figure F3a and b show an example of a filter paper exposed to a rain along with the resultant drop-size distribution.

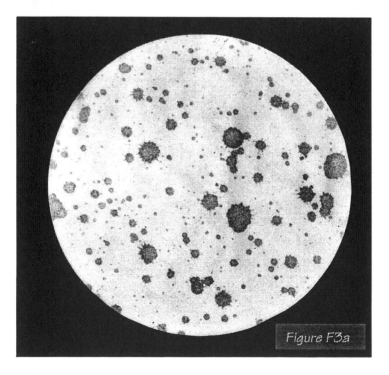

Figure F3a

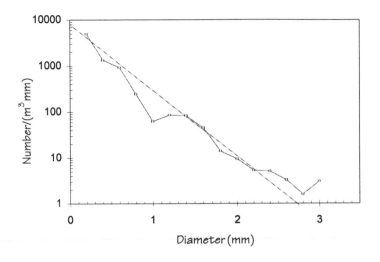

Figure F3. a) Filter paper exposed for 10 s at 0702 CDT, 20 September 1997 in Grand Forks, North Dakota, by Kristi Schueler. The original filter paper had a diameter of 18.5 cm. b) Drop-size distribution from the filter paper. The straight line is a Marshall-Palmer size distribution fit to the data.

Appendix F

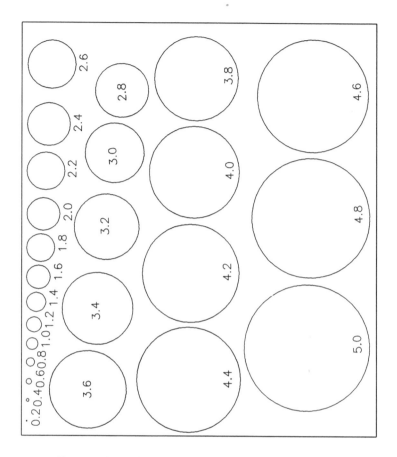

Figure F4   Template for measuring rain drop diameters using filter paper.   This template is based on five empirical and four theoretical calibrations from Radar Meteorology classes at UND from 1987 through 1991.   Drop diameters are in millimeters and the filter paper is Watman No. 1.   The calibration equation is $D_D = 0.479 \, D_S^{0.674}$ where both D_D and D_S are in mm.   To use this template, copy it to clear acetate with 100% magnification, i.e., exactly the same size as shown here.

Glossary

A-scope: A deflection-modulated display in which the vertical deflection is proportional to target echo strength and the horizontal coordinate is proportional to range.

accuracy: Degree of conformity of a measure to a standard or a true value.

A/D converter: Analog-to-digital converter. The electronic device which converts the radar receiver analog (voltage) signal into a number (or count or quanta).

aliasing: The process by which frequencies too high to be analyzed with the given sampling interval appear at a frequency less than the Nyquist frequency.

amplitude: The maximum magnitude of a quantity.

analog: Class of devices in which the output varies continuously as a function of the input.

angular area of sphere = 4π steradians.

anomalous propagation (AP): When nonstandard index-of refraction distributions prevail, "abnormal" or "anomalous" propagation occurs.

Glossary

antenna: A transducer between electromagnetic waves radiated through space and electromagnetic waves contained by a transmission line.

antenna gain: The measure of effectiveness of a directional antenna as compared to an isotropic radiator; maximum value is called antenna gain by convention. Gain can be defined as

$$g = \frac{\textit{power at point with antenna}}{\textit{power at same point with isotropic antenna}}.$$

antenna reflector: The portion of an antenna system which reflects the energy from the radiating element into a focused beam; generally circular parabolas for weather radars.

attenuation: Any process in which the flux density (power) of a beam of energy is dissipated.

autocorrelation: A measure of similarity between displaced and undisplaced (in time, space, etc.) versions of the same function.

automatic gain control: Any method of automatically controlling the gain of a receiver, particularly one that holds the output level constant regardless of the input level.

average power: The average power of a radar is a result of transmitting a high power for a short duration many times a second. Average transmitted power is given by

$$\bar{p}_t = p_t \cdot PRF \cdot \tau$$

where *PRF* is the pulse repetition frequency and τ is the pulse duration.

azimuth: A direction in terms of the 360° compass; north at 0°, east at 90°, south at 180°, west at 270°, etc.

B-scope: An intensity-modulated rectangular display with azimuth angle as the horizontal coordinate and range as the vertical coordinate. Computer printouts of radar parameters in a rectangular display of azimuth versus range is called, by analogy, a B-scan.

backing wind: A change in wind direction in a counterclockwise sense representing cold air advection.

backscatter: That portion of power scattered back in the incident direction

band: *See frequency band.*

bandpass filter: A filter whose frequencies are between given upper and lower cutoff values, while substantially attenuating all frequencies outside these values (this band).

band width: The number of cycles per second between the limits of a frequency band

base data: For NEXRAD, those digital fields of reflectivity, mean radial velocity, and spectrum width data in spherical coordinates provided at the finest resolution available from the radar.

base products: For NEXRAD, those products that present some representation of the base data. This representation may not necessarily be either in full resolution or depict the full area of coverage. Base products can be used to generate a graphic display or further processing.

beam filling: The measure of variation of hydrometeor density throughout the radar sampling volume. The fraction of the radar sample volume filled. If there is no variation in density, the beam is considered to be filled.

beam width: Angular width of antenna pattern. Usually that width where the power density is one-half that on the axis of the beam (half-power or 3-dB point).

bias: A systematic difference between an estimate of and the true value of a parameter.

bin: Radar sample volume

bistatic radar: A radar which uses separate antennas for transmission and reception; usually the transmitter and receiver are at different locations. Bistatic radars depend upon forward scattering of the signal from transmitter to receiver.

boundary layer: The layer of a fluid adjacent to a physical boundary in which the fluid motion is affected by the boundary and has a mean velocity less than the free-stream value.

bounded weak echo region (BWER): A core of weak equivalent reflectivity in a thunderstorm that identifies the location of a strong updraft. The updraft is so strong that large precipitation particles do not have time to form in the lower and mid-levels of the storm and are prevented from falling back into the updraft core from above. The weak echo region is bounded when, in a horizontal section, the weak echo region is completely surrounded or bounded by higher reflectivity values.

bow echo: Rapidly moving, crescent shaped echo that is convex in the direction of motion. Typically associated with strong, straight-line winds.

Bragg scattering: Scatter from small-scale fluctuations (i.e., turbulence) in the refractive index of the atmosphere. Bragg scatter comes from fluctuations which are small compared to the radar's wavelength.

bright band: The enhanced layer of radar echo caused by the difference in radar reflectivity of ice and water particles, the difference in terminal velocity of ice and raindrops, and the sticking and clumping of wet ice particles as they melt. This echo is interpreted as the delineation on a radar display between frozen and liquid precipitation.

CAPPI: Constant Altitude PPI; a data product providing radar data at a fixed height or altitude rather than at a fixed elevation angle.

Cartesian coordinates: The familiar "x-y-z" coordinate system in which the axes are at right angles to each other. The natural coordinate system of radar is polar coordinates, but data are frequently converted to Cartesian coordinates.

Glossary

cell: A compact region of relatively strong vertical air motion (at least 10 m/s).

central limit theorem: Statistical theorem showing that averages approach a Gaussian distribution independent of the input distribution.

centroid: The center of mass of a storm.

clutter: Echoes that interfere with observation of desired signals on a radar display. Usually applied to ground targets.

coherent radar: A radar that utilizes both signal phase and amplitude to determine target characteristics

COHO: Coherent oscillator.

cokriging: A technique for estimating values of a spatial process (e.g., a precipitation field) given point observations of the process (e.g., rain gage observations) and possibly auxiliary observations (e.g., radar and satellite observations).

complex index of refraction: $m = n + i\,k$, where n is normal index of refraction, $i = \sqrt{-1}$, and k is absorption coefficient.

complex signal: A signal containing both amplitude and phase information.

conjugate of complex number: If $c = a + ib$ is a complex number, then $c^* = a - ib$ is its complex conjugate.

convergence: A measure of the contraction of a vector field.

correlation: A measure of similarity between variables or functions.

couplet: Adjacent maxima of radial velocities of opposite signs.

covariance: A measure of the degree of association between two variables. In Doppler radars, the argument (or angle) of the covariance of the complex signal is a measure of the Doppler frequency.

cross section: *See radar cross section.*

CRT: Cathode ray tube.

curvature: The reciprocal of the radius of a circle; the rate of change in the deviation of a given arc from any tangent to it.

CW: Continuous wave

data resolution: For NEXRAD, the resolution of the base data as produced by the signal processor, nominally 1° in azimuth by 1° in elevation by 0.54 n.mi (1 km) for reflectivity and by 0.13 n.mi (0.25 km) for velocity values.

dBm: The logarithmic unit of absolute power referenced to a power of 1 mW.

dBZ: The logarithmic unit of radar reflectivity factor defined as

$$Z = 10 Log_{10}\left(\frac{z}{1mm^6 / m^3}\right)$$

where Z is the logarithmic and z is the linear radar reflectivity factor, respectively.

dealiasing: Process of correcting for aliases in the velocity measurement.

decibel (dB): A logarithmic unit used to express the ratio of two quantities. Mathematically,

$$power\ ratio = 10\ Log_{10}\left(\frac{P_1}{P_2}\right).$$

dielectric constant: For a given substance, the ratio of the capacity of a condenser with that substance as dielectric to the capacity of that condenser with a vacuum as dielectric.

dielectric material: A substance that contains no or few free charges and which can support electromagnetic stress.

differential reflectivity: A measure of the difference in radar reflectivity factor of a target measured using horizontal and vertical polarization. Mathematically,

$$Z_{DR} = 10 Log_{10}\left(\frac{z_H}{z_V}\right)$$

where z_H and z_V are the linear radar reflectivity factors at horizontal and vertical polarization, respectively. Z_{DR} is measured in decibels.

diffraction: The process by which the direction of radiation is changed so that it spreads into the geometric shadow region of an opaque or refractive object that lies in a radiation field.

disdrometer: Equipment that measures and records the size distribution of raindrops.

distortion: Change in a signal resulting in gross non-linearities in signal processing or handling.

divergence: A measure of the expansion in a vector field.

Doppler dilemma:

$$V_{max} R_{max} = \frac{c \lambda}{8}$$

where V_{max} is maximum unambiguous velocity, $Rmax$ is maximum unambiguous range, c is speed of light, λ is wavelength. This term coined first by Rodger Brown, NSSL.

Doppler frequency shift:

$$f_d = \frac{2V}{\lambda}$$

where V is radial velocity of target, λ is wavelength.

Doppler shift: The change in frequency at a receiver due to the relative motion of the receiver and the energy source.

downburst: A strong downdraft that induces an outburst of damaging winds on or near the ground.

Glossary

downdraft: Current(s) of air with marked vertical downward motion.

dryline: A mesoscale feature with its own associated vertical circulation. It is a narrow, almost vertical zone, across which a sharp moisture gradient, but little temperature gradient, occurs at the Earth's surface.

dual Doppler: The use of two Doppler radars to measure two different radial velocities and combine them into a two- or three-dimensional flow field, depending upon the boundary conditions and assumptions used in the processing.

ducting: The phenomenon by which the radar signal propagates along the boundary of two dissimilar air masses. The range to which ground echoes are detected during ducting conditions are greatly extended; gaps in coverage can also appear. Ducting occurs when the upper air is exceptionally warm and dry in comparison with the air at the surface. Ducting occurs when

$$dN/dh \leq -157 \text{ N-units/km.}$$

Where N is refractivity and h is height in km.

duplexer: A device in the waveguide which protects the sensitive receiver from the full power of the transmitter; usually contains one or more TR (transmit-receive) tubes.

dwell time: Time over which a signal estimate is made. Usually, the time required for the antenna to traverse one degree.

dynamic range: The ratio, usually expressed in decibels, of the maximum to the minimum signal that a system can handle. Used to describe limits of receivers.

echo: Energy backscattered from a target as seen on the radar display.

effective radar reflectivity factor: Same as equivalent radar reflectivity factor.

elevation angle: The vertical pointing angle of the antenna above the horizon (the WSR-88D antenna can vary from -1° to +60°; most radars can scan to 90° elevation angles).

equivalent radar reflectivity factor z_e or Z_e: The concentration of uniformly distributed small (diameter one sixteenth wavelength or less) water particles that would return the amount of power received. In this text, this is expressed in mm^6/m^3 for linear values of equivalent radar reflectivity factor z_e and in dBZ for logarithmic values of equivalent radar reflectivity factor Z_e.

estimate: A statement of the value of a quantity or function based on a finite number of samples.

eye wall: The area of tall cumulonimbus storms surrounding the eye of a hurricane. Heavy rain and very high winds occur in the eye wall. The area inside the eye wall is a roughly circular area of comparatively light winds and fair weather found at the center of a severe tropical cyclone or hurricane.

Glossary

feeder cloud: The flanking lines of developing cumulus congestus clouds that sometimes merge with and appear to intensify supercells.

folding: Aliasing; applied to both velocity and range aliasing.

fractional uncertainty $\delta q / q$: The uncertainty δq in the measurement of a quantity q expressed as a fraction. Fractional uncertainty is a unitless parameter. To express uncertainty in percent, multiply by 100%, e.g.,

$$\% \ uncertainty \ = \ \frac{\delta q}{|q|} \bullet 100\%.$$

Absolute signs are used around q to insure the expression of uncertainty is always positive.

freezing level: The lowest altitude in the atmosphere over a given location where the air temperature is 0°C.

frequency: The number of recurrences of a periodic phenomenon per unit time. Electromagnetic energy is usually specified in Hertz (Hz), which is a unit of frequency equal to one cycle per second. Weather radars typically operate in the gigahertz (GHz) range. *See wavelength.*

frequency band: A range of frequencies between some upper and lower limits.

frequency carrier: The fundamental transmitted microwave frequency of a transmitter. It is modulated so that it exists for a few microseconds each pulse. For the WSR-88D, the fundamental transmitted microwave frequency is between 2,700 and 3,000 MHz.

Fresnel reflection: The reflection of a radar signal from a single, dominating discontinuity of the refractive index, usually with a large horizontal extent. Also called "partial reflection" because only a small fraction of the incident power is reflected, "specular reflection" if the horizontal surface discontinuity is assumed to be smooth, or "diffuse reflection" if the discontinuity is assume to be corrugated or somewhat rough.

Fresnel scatter: Scatter which occurs if several or many refractive index discontinuities exist along the pointing direction. The difference between Fresnel reflection and scatter may be primarily a matter of resolution of the sampling volume compared to the size of the reflecting target.

gain: A change in signal power, voltage or current. Usually applied to a change greater than one and frequently expressed in decibels. *See: antenna gain.*

gating (range gating): The use of electric circuits in radar to eliminate or discard the target signals from all targets falling outside certain desired range limits.

Gaussian: Refers to the normal distribution; phenomena whose events are "normally" distributed are "Gaussian" distributed. This is the most common distribution encountered in physical processes.

Glossary

GPS: Global Positioning System. A network of satellites which provide extremely accurate position and time information. Useful in remote locations or for moving platforms.

graupel: A rimed ice aggregate often found in vigorous storms. It is formed when ice aggregate collects supercooled liquid water droplets.

ground clutter: The pattern of radar echoes from fixed ground targets.

gust front: The boundary between the horizontally propagating cold air outflow from a thunderstorm and the surrounding air.

gyro: A device used for measuring changes in direction. Sometimes used for automatic antenna stabilization for moving radars.

hail: Precipitation in the form of spherical or irregular ice produced by convective clouds, usually cumulonimbus. By convention, hail has a diameter of 5 mm or more. Smaller particles of similar origin may be classified as ice or snow pellets.

hertz (Hz): A unit of frequency equal to one cycle per second.

homodyning: The transfer of signal intelligence from one carrier to another by mixing of signals at different frequencies.

hook echo: A pendant or hook on the right rear of an echo that often identifies mesocyclones on the radar display. The hook is caused by precipitation drawn into a cyclonic spiral by the winds, and the associated notch in the echo is caused by precipitation-free, warm, moist air flowing into the storm.

hydrometeor: A particle of condensed water (liquid, snow, ice, graupel, hail) in the atmosphere.

i: $i = \sqrt{-1}$; a mathematical operator which, when multiplied with a number or parameter, has the effect of turning the vector 90° counter clockwise from its original position.

I: *See in-phase.*

incident power density: Energy per unit area incident on the radar target.

index of refraction: *See refractive index.*

in-phase: The component of a complex signal along the real axis in the complex plane.

INS: Inertial Navigation System (or Unit): A highly accurate tool for measuring and keeping track of motions and accelerations. Often composed of laser gyros. Can be used in stabilizing antennas on moving platforms.

isodop: Contour of constant Doppler velocity values.

isolated storm: An individual cell or a group of cells that are identifiable and separate from other cells in a geographic area.

Glossary

isotropic: Having the same characteristics in all directions, as with isotropic antennas. Directional or focused antennas are *not* isotropic.

$|K|^2$: $K = \dfrac{m^2 - 1}{m^2 + 1}$ where $|K|$ is the magnitude of the expression for the complex index of refraction m. For water, $|K|^2 = 0.93$; for ice $|K|^2 = 0.197$.

Kalman filter: A linear system in which the mean squared error between the desired output and the actual output is minimized when the input is a raondom signal generated by white noise.

klystron: An electron tube used as a low-power oscillator or a high-power amplifier at ultrahigh frequencies. Noted for exceptional stability over long periods of transmission.

life on a logarithmic scale (*see* Appendix A for details):

Logarithmic scale:	0	1	2	3	4	5	6	7	8	9	10
Linear scale:											
factors of 2 & up:	1		2		4		8				
factors of 2 & down:	1.25		2.5		5						10
± factors of 2:		1.5		π		6					

line echo wave pattern (LEWP): A radar echo pattern formed when a segment of a line of thunderstorms surges forward at an accelerated rate. A meso-high pressure area is usually present behind the accelerating thunderstorms. A meso-low pressure area is usually present at the crest of the wave.

look angle: A given radar's perspective on a storm; i.e., the angle at which its antenna beam hits it. By using more than one radar with different look angles, multiple Doppler analyses can be performed to retrieve two- or three-dimensional winds.

macroburst: Large downburst with 4 km or larger outflow size with damaging wind lasting 5 to 20 minutes.

mainlobe: The envelope of electromagnetic energy along the main axis of the beam.

magnetron: A self-exciting oscillator tube used to produce the radio frequency signal transmitted by some radars. It utilizes a strong magnetic field to help induce the RF signal generated.

maximum unambiguous range: The maximum range to which a transmitted pulse wave can travel and return to the radar before the next pulse is transmitted.

$$R_{max} = \frac{c}{2PRF}$$

where c is speed of light, PRF is pulse repetition frequency (Hz).

maximum unambiguous velocity: The maximum velocity which a Doppler radar can determine unambiguously.

$$V_{max} = \frac{PRF\lambda}{4}$$

where PRF is pulse repetition frequency (Hz), λ is wavelength.

mean Doppler velocity: Reflectivity-weighted average velocity of targets in a given volume sample. Usually determined from a large number of successive pulses. Also called mean radial velocity. Doppler velocity usually refers to spectral density first moment; radial velocity to base data.

melting level: That height in the atmosphere at which ice melts to water, i.e., at which the temperature is 0°C. This may range from 0 to 5 km above the earth's surface.

mesocyclone: A 3-dimensional region in a storm that rotates cyclonically and is closely correlated with severe weather.

mesoscale convective complex (MCC): A quasi-circular conglomeration of thunderstorms having a cloud-top area larger than 100,000 km² and persisting for more than 6 hours.

mesoscale convective system (MCS): Precipitation systems 20 to 500 km wide that contain deep convection. Examples in mid-latitudes are large isolated thunderstorms, squall lines, Mesoscale Convective Complexes, and rainbands.

meteorological radar equation:

$$p_r = \frac{\pi^3 \, p_t \, g^2 \, \theta \, \phi \, h \, |K|^2 \, z}{1024 \, \ln 2 \, \lambda^2 \, r^2}$$

where p_r is received power, p_t is transmitted power, g is antenna gain, λ is wavelength, r is range to target, θ and ϕ are horizontal & vertical beamwidths, h is the effective pulse length (i.e., $h = c\tau/2$ where c is the speed of light and τ is the pulse duration; the factor of 2 is required to insure energy from the pulse volume arrives back at the receiver simultaneously), $|K|^2$ is dielectric constant term, z is radar reflectivity factor; all terms are linear, not logarithmic.

microburst: Small downburst, ≤ 4 km in outflow size (i.e., ≤ 2.5 n mi), with peak winds lasting 2 to 15 min.

Mie scattering or resonant region: Radar backscattering by targets having dimensions somewhat greater than 1/10 the wavelength of the radar but less than several radar wavelengths.

modulation: Variation of the amplitude, frequency, or phase of a wave due to the mixing of two signals.

monostatic radar: A radar that uses a common antenna for both transmitting and receiving.

multicell storm: A storm that consists of a cluster of single cells that are often short lived.

multiple Doppler analysis: The use of more than one radar (hence, more than one look angle) to reconstruct spatial distributions of the two- or three-dimensional wind field. This cannot be measured from a single radar. It includes dual-, triple- and over-determined multiple-Doppler analyses.

Glossary

NEXRAD: NEXt generation RADar; WSR-88D system.

NEXRAD base data: Those digital fields of reflectivity, mean radial velocity and spectrum width data in spherical coordinates provided at the finest resolution available

notch width: The 3 dB band width of a rejection filter.

Nyquist frequency: The highest frequency that can be determined in data that have been discretely sampled. For data sampled at frequency f_s, this frequency is $f_s/2$. Doppler radar sampling rate is equal to the pulse repetition frequency (*PRF*).

Nyquist interval (Nyquist velocity): The maximum unambiguous velocity that can be measured by a Doppler radar.

Nyquist sampling theorem: In order to unambiguously measure a frequency, a sampling rate of at least two times this frequency is required. Doppler radar sampling rate is equal to the pulse repetition frequency (*PRF*).

Ohm's law: $I = E/R$ where I is current (amperes), E is electromotive force (i.e., voltage), and R is resistance (ohms).

oscillator: The general term for an electric device that generates alternating currents or voltages. The oscillator is classified according to frequency of the generated signal.

over-determined multiple-Doppler analysis: A multiple Doppler analysis technique in which there are more radars than the desired number of wind components to be retrieved. The wind retrieval can then be performed as an "optimization" to reduce the retrieval errors.

overhang: A storm has overhang if the edge of the storm at a given mid-level height extends outward beyond the edge of the storm at bottom.

parabolic antenna: An antenna with a radiating element and a parabolic reflector that concentrates the radiated power into a beam; it also concentrates the returned signal.

peak power: The amount of power transmitted by a radar during a given pulse. Note that because these pulses are widely spaced, the average power will be much smaller.

pedestal: A generic term for the structure supporting the antenna dish. Usually includes the drive motors and one end of the servo loop.

persistence: On a radar display, the length of time during which a signal is visible; modern color displays are no longer concerned with persistence.

phase: A particular angular stage or point of advancement in a cycle; the fractional part of the angular period through which the wave has advanced, measured from the phase reference.

phase shift: The angular difference of two periodic functions.

Glossary

phasor diagram: A diagram used to represent complex numbers. The x-axis is the real component and the y axis is the imaginary component. The x axis can be the in-phase and the y axis the quadrature components (I and Q components).

planetary boundary layer: The layer of the atmosphere from the earth's surface to the geostrophic wind level (about 1 to 1.6 km above the surface). Above this level lies the free atmosphere where the frictional influence of the earth's surface on air motion is negligible.

platform: A generic term often used to encompass the pedestal and antenna assembly; sometimes including the radar control, display and analysis hardware and software as well.

point target radar equation:

$$p_r = \frac{p_t \, g^2 \, \lambda^2 \, \sigma}{64 \, \pi^3 \, r^4}$$

where p_r is received power, p_t is transmitted power, g is antenna gain, λ is wavelength, r is range to target, σ is backscattering cross-sectional area; all terms are linear, not logarithmic.

polar coordinates: *See* spherical coordinates.

polarization radar: A radar which uses polarization information and how it affects and is affected by the signal backscattered from a target. Radars which alternately transmit horizontal and vertical polarizations are used to measure differential reflectivity.

power: $P = I V = V^2 / R = I^2 R$ where I is current (amps), V is voltage (volts), R is resistance (ohms) and P is power (watts)

power ratio (linear): power ratio $= p_1/p_2$ where p_1 and p_2 are two powers measured in the same units.

power ratio (logarithmic): Power Ratio $= 10 \log_{10}(p_1/p_2)$, in decibels, where p_1 and p_2 are two powers measured in the same units; logarithmic power: P (dBm) $= 10 \log_{10}[p_1/(1\ mW)]$ where p_1 is a power measured in mW.

PPI (plan-position indicator): An intensity-modulated display on which echo signals are show in plan position with range and azimuth angle displayed in polar coordinates, forming a map-like display.

precision: The accuracy with which a number can be represented, i.e., the number of digits used to represent a number.

propagation: Transmission of electromagnetic energy as waves through or along a medium.

PRT: Pulse repetition time; PRT $= 1/PRF$.

pulse: A single, short-duration transmission of electromagnetic energy.

pulse duration: Time occupied by a burst of transmitted radio energy. This may also be expressed in units of range (pulse length) or time (pulse duration). Also called pulse width.

pulse length *h*: $h = c\,\tau$ where τ is duration of transmitted pulse, *c* is speed of light, *h* is length of pulse in space. However, in the radar equation, the length *h*/2 is actually used for calculating pulse volume because we are only interested in signals that arrive back at the radar simultaneously.

pulse-pair processing: Nickname for the technique of mean velocity estimation by calculation of the signal complex covariance argument. The calculation requires two consecutive pulses.

pulse radar (or pulsed radar): A type of radar designed to facilitate range measurement in which the transmitted energy is emitted in periodic, brief transmissions.

pulse repetition frequency (*PRF*): The number of pulses transmitted per second. Also called pulse repetition rate.

pulse repetition time (*PRT*): The time interval from the beginning of one pulse to the beginning of the next succeeding pulse.

pulse resolution volume: A discrete radar sampling volume. The size is determined by the horizontal and vertical beamwidths and the pulse duration or pulse length.

pulse width: The time occupied by an individual broadcast from a radar.

PUP: Principal User Processor.

Q *or* **quadrature:** The component of the complex signal that is 90° out of phase with the in-phase component. This component lies along the imaginary axis in the complex plane.

radar cross section: The area of a fictitious, perfect reflector of electromagnetic waves that would reflect the same amount of energy back to the radar as the actual target.

radar reflectivity η: $\eta = \Sigma(\sigma)/unit\ volume$ where σ is backscattering cross-sectional area and the summation is done over a unit volume. η and z are related through

$$\eta = \frac{\pi^5 |K|^2 z}{\lambda^4}$$

where $|K|^2$ is the dielectric constant term usually taken to be 0.93, λ is wavelength, and z is radar reflectivity factor. η is usually expressed in units of cm^{-1} while z is usually expressed in units of mm^6/m^3; consequently, a unit conversion factor is also required to convert from one to the other.

radar reflectivity factor z: $z = \Sigma(N_i\ D_i^6)$ where N_i is number of drops of diameter D_i per unit volume; units of z are mm^6/m^3. [Note that many radar meteorologists use capital Z for both linear and logarithmic units while others use ζ for logarithmic reflectivity and z for linear reflectivity; this text uses z for linear radar reflectivity factor and Z for logarithmic radar reflectivity factor. The same lower-case/capital-letter convention is generally used herein with all parameters which are frequently expressed in both linear and logarithmic units.]

radial velocity (V_r): The component of motion of the target toward or away from the radar. Doppler radars only detect radial components of velocity.

random variable (variate): A variable characterized by random behavior in assuming its different possible values. Mathematically, it is described by its probability distribution, which specifies the possible values of a random variable together with the probability associated (in an appropriate sense) with each value. A random variable is said to be "continuous" if its possible values extend over a continuum, "discrete" if its possible values are separated by finite intervals.

range: Distance from the radar antenna to a target (based on round-trip time).

range bin: A sampling point at which reflectivity, velocity, and/or spectrum-width data are collected. Typical radars have as many as 1000 range bins or range gates along a single radial. Typically, 10 to 100 individual pulses are averaged together to get a single point value.

range folding: Apparent range placement of a multiple-trip return. A multiple-trip return appears at the difference of the true range and a multiple of the unambiguous range, i.e.,

$$R_{displayed} = R_{true} - n\, R_{max}$$

where R_{true} is the true range of the echo, $R_{displayed}$ is the range where the echo is displayed, and n = 0, 1, 2,....

range gate: The discrete point in range along a single radial of radar data at which the received signal is sampled. Range gates are typically spaced at 100-m intervals or larger.

range unfolding: Process of removing range ambiguity in apparent range of a multitrip target.

Rankine vortex: Velocity profile for a symmetric circulation in which the inner core is in solid rotation and tangential winds outside the core vary inversely with radial distance from the center.

Rayleigh scattering: Scattering by spherical particles whose radii are smaller than about one-tenth the radar wavelength.

RDA: Radar Data Acquisition system in WSR-88D radar system.

receiver: The electronic device which detects the backscattered radiation, amplifies it and converts it to a low-frequency signal which is related to the properties of the target.

reflectivity: A measure of the fraction of radiation reflected by a given surface; defined as a ratio of the radiant energy reflected to the total that is incident upon that surface. Lazy radar meteorologists and others working with radar data from storms frequently say "reflectivity" when they should say radar reflectivity factor or equivalent radar reflectivity factor; forgive us.

refraction: The process in which the direction of energy propagation is changed as a result of a change in the speed of propagation caused by changes in density within the medium or as the energy passes through the interface representing a density discontinuity between two media.

refractive index: A measure of the amount of refraction. Numerically equal to the ratio of wave velocity in a vacuum to wave speed in the medium, i.e., $n = c/u$ where u is actual speed, and c is speed of light in a vacuum.

refractivity N: $N = (n-1)10^6$ where n is refractive index and

$$N = \frac{77.6}{T}\left(P + \frac{4810e}{T}\right)$$

where T is absolute temperature in K, P is pressure in hPa, and e is vapor pressure in hPa.

resolution: The smallest increment of a measurement of a parameter.

rf: radio frequency.

RHI (range-height indicator): An intensity-modulated display with height as the vertical axis and range as the horizontal axis. A cross-section in a vertical plane passing through the radar.

Rinehart projection: A map projection which shows where range-aliased echoes are in relationship to ground features. It is generated by shifting all map points in toward the radar by exactly R_{max} (for a "second-trip" map; for a "third-trip" echo map, all points would be shifted by $2 \cdot R_{max}$) where R_{max} is the maximum unambiguous range of the radar.

RPG: Radar Products Generator system in WSR-88D radar system.

sample and hold: The process of sampling (measuring) the signal strength at a particular point in space (i.e., at a range gate).

sea breeze: A current of air flowing inland and associated with warmer surface temperatures inland than at sea. Often shows as a long, thin radar feature. Frequently attributed to insects being caught up in the frontal region. Temperature and moisture gradients across the front may also contribute to its reflectivity.

sectorized hybrid scan: A single reflectivity scan composed of data from the lowest four elevation scans. Close to the radar, higher tilts are used to reduce clutter. At further ranges, either the maximum values from the lowest two scans are used or the second scan values are used alone.

servo loop: In radar meteorology, a generic description of hardware needed to remotely control the motion of the antenna dish.

severe thunderstorm: A storm with a tornado, surface hail $\geq 3/4$ inch, or wind gusts ≥ 50 knots, or all three.

shear: The rate of change of the vector wind in a specified direction normal to the wind direction.

Vertical shear is the variation of the horizontal wind in the vertical direction.

shelf cloud: A type of arcus (or roll) cloud. It is a low-level, horizontal, accessory cloud that appears to be wedge shaped as seen along the leading edge of approaching thunderstorms. It is accompanied by gusty, straight-line winds and is followed by precipitation.

sidelobe: Secondary radiated energy maximum in a direction other than that of the radar main lobe. Typically contains a small percent of energy compared to the mainlobe but may produce erroneous echoes.

signal to noise ratio (SNR): A ratio that measures the comprehensibility of data, usually expressed as the signal power divided by the noise power, usually expressed in dB.

spearhead echo: A radar echo associated with a downburst with a pointed appendage extending toward the direction of the echo motion. The appendage moves much faster than the parent echo, which is drawn into the appendage. During the mature stage, the appendage turns into a major echo and the parent echo loses its identity

specific humidity: In a system of moist air, the ratio of the mass of water vapor to the total mass of the system; measured in units of g/kg.

spectral density: The distribution of power by frequency.

spectrum width σ: A measure of dispersion of velocities *within* the radar sample volume. Standard deviation of the velocity spectrum. Spectrum width from meteorological targets is given by

$$\sigma^2 = \sigma_s^2 + \sigma_d^2 + \sigma_a^2 + \sigma_t^2$$

where subscripts mean: s is wind shear (e.g., (m/s)/km), d is drop terminal velocity term, a is antenna scan rate term, t is turbulence term.

sphere calibration: An incorrect term which refers to measuring the gain of a radar's antenna by using a metal sphere of known back-scattering cross-sectional area. The sphere is often tethered to a balloon so it is easier to detect. This term is "incorrect" because the sphere is not calibrated; the gain of the antenna is measured.

spherical coordinates: A coordinate system which uses range, azimuth and elevation angle to locate positions in space relative to a point. The natural coordinate system of a radar. Sometimes referred to as polar coordinates although mathematicians would say polar coordinates are range, azimuth, and height.

squall line: A line or narrow band of active thunderstorms.

STALO: Stable local oscillator.

standard deviation: The positive square root of the variance of a parameter. The velocity standard deviation is called spectrum width.

standard refraction: Refraction in the atmosphere when temperature and humidity distributions are approximately average. For standard refraction, $\delta N/\delta h = -39.2$ N-units/km where δN is the change in refractivity over height interval δh.

storm: Any disturbed state of the atmosphere, especially as affecting the Earth's surface, and strongly implying destructive and otherwise unpleasant weather. Storms range in scale from tornadoes and thunderstorms through tropical cyclones to widespread extratropical cyclones.

storm motion: The velocity at which a storm travels.

subrefraction: A condition of atmospheric refraction when radar waves are bent less than normal. Subrefraction occurs when $\delta N/\delta h > -39.2$ N-units/km; e.g., -30 or -20 N-units/km.

sun pointing: Alignment of the radar antenna by locating the position of the sun in the sky. Solar position is predictable to high precision, so this is a good check on azimuth and elevation alignment of the antenna. Solar flux density is measured by observatories around the world, so measurements of the strength of the solar signal can also be used to check receiver sensitivity and antenna gain.

supercell: A large, long-lived (up to several hours) cell consisting of one quasi-steady updraft-downdraft couplet that is generally capable of producing the most severe weather (tornadoes, high winds, and giant hail).

supercooled liquid water: Water droplets in the atmosphere that are colder than 0°C. Supercooled liquid water can exist at temperatures well below 0°C. Important for the growth of graupel and hail. A hazard to aviators.

superrefraction: A condition of atmospheric refraction when radar waves are bent more than normal. Superrefraction occurs when $\delta N/\delta h < -39.2$ N-units/km, e.g., -50 or -60 N-units/km.

synchronous detection: Processing that retains the received signal amplitude and phase but that removes the intermediate frequency carrier.

target: Precipitation or other phenomena that produce echoes.

TDWR: Terminal Doppler Weather Radar.

thin line echo: A narrow, elongated non-precipitating echo usually associated with thunderstorm outflow, fronts, or other density discontinuities; also know as a fine line.

tilt: The elevation angle used by a radar antenna. Another name for a PPI scan; a conical scan at a given elevation angle.

time-height display: An intensity-modulated display which has height as the vertical coordinate and time as the horizontal coordinate; usually used for vertically-pointing antennas only.

tornado vortex signature (TVS): The radar "signature" of a vortex indicative of a tornado or tornadic circulation. A small-scale anomalous region of high shear associated with a tornado.

transmitter: The equipment used for generating and amplifying a radio frequency (RF) carrier signal, modulating the carrier signal with intelligence, and feeding the modulated carrier to an antenna for radiation into space as electromagnetic waves. Weather radar transmitters are usually magnetrons or klystrons.

triple Doppler: The use of three Doppler radars, each of which scans a storm from a different direction. Since any wind has three components, a single radar can measure only one (radial) direction. If three different radars are used, the three measured radial velocities can be transformed into the three-dimensional velocity field. Since hydrometeors have a terminal velocity, the velocity field will not be the wind field unless some other information or assumption is also used.

unambiguous range: The range to which a transmitted pulse wave can travel and return to the radar before the next pulse is transmitted.

uncertainty: An estimate of the degree to which the experimental measurement of a parameter differs from the true value. The uncertainty of a quantity q is given by δq. Uncertainty δq given in this form has the same units as the parameter q. Fractional uncertainty $(\delta q / q)$, however, is a unitless quantity.

uncertainty for $q = x + y + z$: For additive parameters the upper limit uncertainty is given by

$$\delta q = \delta x + \delta y + \delta z;$$

for random, independent uncertainties,

$$\delta q = (\delta x^2 + \delta y^2 + \delta z^2)^{0.5}$$

uncertainty for $q = \dfrac{xy}{uv}$: For random, independent uncertainties:

$$\frac{\delta q}{|q|} = \left(\left(\frac{\delta x}{|x|}\right)^2 + \left(\frac{\delta y}{|y|}\right)^2 + \left(\frac{\delta u}{|u|}\right)^2 + \left(\frac{\delta v}{|v|}\right)^2 \right)^{0.5}.$$

uncertainties for $q = B\,x$: $\delta q = |B|\,\delta x$.

uncertainties for $q = x^p$: $\dfrac{\delta q}{|q|} = p\,\dfrac{\delta x}{|x|}$.

uncertainties in logarithmic parameter Z:

$$dZ \cong 4\frac{\delta z}{z}$$

where z is linear radar reflectivity factor (mm^6/m^3) and δZ is logarithmic value of uncertainty in radar reflectivity factor (dB). This relationship can be applied to any pair of linear and logarithmic parameters.

uncertainty in linear reflectivity factor z:

$$\frac{\delta z}{z} \cong \frac{\delta Z}{4}$$

where δZ is the logarithmic uncertainty of radar reflectivity factor (dB), and z is linear radar reflectivity factor (mm^6/m^3).

unimodal: A distribution having only one localized maximum, i.e., only one peak.

updraft: Current(s) of air with marked vertical upward motion.

unimodal: A distribution having only one localized maximum, i.e., only one peak.

UTC: Universal Coordinated Time.

VAD: Velocity-azimuth display; now VAD is used to indicate a display of winds with height and time derived from a VAD analysis of the radial winds as a function of height.

vapor pressure e: The partial pressure contributed by the presence of water vapor in the atmosphere. Vapor pressure can be calculated from the dew-point temperature using

$$e_s = 6.1121 \exp\left(\frac{\left(18.729 - \frac{T}{227.3}\right)T}{T + 257.87}\right)$$

where e_s is the saturation vapor pressure (hPa) at temperature T (°C); alternatively, e can be the actual vapor pressure if T is the dew-point temperature (°C).

variance: A measure of variability.

veering wind: A change in wind direction in a clockwise sense representing warm air advection.

velocity aliasing: Ambiguous detection of radial velocities outside the Nyquist interval. Variance will be high where the radial velocity display suddenly switches from one extreme to the other (i.e., where velocity aliasing is taking place).

VIL: Vertically integrated liquid water content. A measure of the total amount of water above a point. Measured in kg/m^2. One equation to calculate VIL is

$$VIL = 3.44 \cdot 10^{-6} \sum z^{4/7} \Delta H$$

where z is the linear radar reflectivity factor (mm^6/m^3) and H is height in meters.

vortex: In its most general use, any flow possessing vorticity. More often the term refers to a flow with closed streamlines.

vorticity: A vector measure of local rotation in a fluid flow.

VVP: Volume Velocity Processing. A way to estimate large-scale two-dimensional winds, divergence, and fall speeds from one-dimensional radial-velocity data. Essentially a multivariate regression which fits a simple wind model to the observed radial velocities. Very similar to VAD and EVAD except that it uses different functions for the fit.

wall cloud: A local, abrupt lowering of a rain-free cumulonimbus base into a low-hanging accessory cloud, from 2 to 6 km (1 to 4 miles) in diameter. The wall cloud is usually located in the southwestern part of a severe thunderstorm in the main updraft to the southwest of the main precipitation region. Rapid upward motion and visible rotation may be seen in wall clouds from several km away. Almost all strong tornadoes develop from wall clouds.

watershed: The total area drained by a river and its tributaries.

watt: The unit of power in the meter-kilogram-second (mks) system of units; equal to one joule per second.

waveguide: A hollow conductor, usually rectangular or round in cross-section, used to carry radar waves between various components of a radar.

wavelength: The distance a wave will travel in the time required to generate one cycle. The distance between two consecutive wave peaks (or other reference points) in space. Wavelength λ can be expressed as $\lambda = c/f$ where c is the speed of light and f is the frequency.

weak echo region (WER): Within a convective echo, a localized minimum of equivalent reflectivity associated with the strong updraft region.

WSR-88D system: The summation of all hardware, software, facilities, communications, logistics, staffing, training, operations, and procedures specifically associated with the collection, processing, analysis, dissemination, and application of data from the WSR-88D unit.

WSR-88D unit: The combination of one RDA, one RPG, and all associated RPGOPs and PUPs, and interconnecting communications. "WSR" is Weather Surveillance Radar; commissioned in 1988; "D" is Doppler capability.

Z-R **relationship:** An empirical relationship between radar reflectivity factor z (in mm^6/m^3) and rainrate R (in mm/h), usually expressed as $z = A\, R^b$ where A and b are empirical constants; e.g., $z = 300\, R^{1.6}$.

Constants and Conversion Factors

speed of light $c =$	$299\ 792\ 456 \pm 10$ m/s
Boltzmann's constant $\sigma =$	$5.66961 \pm 96 \cdot 10^{-5} \dfrac{erg}{cm^2\ s\ K^4}$
Boltzmann's constant $k =$	$1.380622 \pm 59 \cdot 10^{-16}$ erg/K
1 in $=$	25.4 mm
1 ft $=$	.3048 m
1 n mi $=$	1852 m
1 MW $=$	1000 kW
1 kW $=$	1000 W
1 W $=$	1000 mW
1 mW $=$	1000 μW
$1° =$	$\pi/180°$
1 radian $=$	$180°/\pi$

Note: Except for the first three constants, all of the above conversion factors or constants are exactly correct as given, some by international definition.

References

AMS, 1952: *Compendium of Meteorology*, Amer. Meteor. Soc., Boston.

Atlas, David, 1964: Advances in radar meteorology. *Adv. Geophys.*, **10**, Landsberg and Mieghem, Eds., Academic Press, 317-478.

Atlas, David, Editor, 1990: *Radar in Meteorology*, Boston, Amer. Meteor. Soc., 806 pp.

Barge, Brian L., 1970: Polarization observations in Alberta, Preprints, *14th Radar Meteor. Conf.*, Tucson, Arizona, Amer. Meteor. Soc., Boston, 221-224.

Battan, L. J., 1959: *Radar Meteorology*, Chicago, University of Chicago Press, 161 pp.

Battan, L. J., 1973: *Radar Observation of the Atmosphere.* Chicago, University of Chicago Press, 324 pp. [Reprinted by: TechBooks, 2600 Seskey Glen Court, Herndon, VA 22071, phone 703-758-1518]

Bean, B. R., and E. J. Dutton, 1968: *Radio Meteorology*, Dover Publications, Inc., New York, 435 pp.

Beard, K., 1985: Simple altitude adjustment to raindrop velocities for Doppler radar analysis, *J. Atmos. & Ocean. Tech.*, 2, 468-471.

References

Buderi, Robert, 1996: *The Invention That Changed the World*. Simon & Schuster, New York, 575 pp.

Castelli, J. P., W. R. Barron and J. Aarons, 1975: An atlas of quiet sun radio frequency measurements made at Sagamore Hill Solar Radio Observatory, 1966-1974. Air Force Cambridge Research Laboratories, AFCRL-TR-75-0132 (Special Reports No. 189).

Clark, Alistair, 1997: The Radar Entomology Web Site. http://www.adfa.oz.au/~vad/trews/ww_re_hp.htm

Cohen, E. Richard, and Barry N. Taylor, 1987: The 1986 CODATA recommended values of the fundamental physical constants, *J. Res. Of the Nat. Bureau of Standards*, **92**, 85. [http://physics.nist.gov/PhysRef Data/codata86.bk/indexcodata.html]

Collier, C. G., 1989: *Applications of Weather Radar Systems, A Guide to Uses of Radar Data in Meteorology and Hydrology*. John Wiley & Sons, New York, 294 pp.

Donaldson, R. J., Jr., 1964: A demonstration of antenna beam errors in radar reflectivity patterns. *J. Appl. Meteor.*, **3**, 611-623.

Doviak, Richard J., and Dusan S. Zrnic', 1993: *Doppler Radar and Weather Observations, Second Edition*. San Diego, Academic Press, Inc., 562 pp.

Eccles, Peter, 1975: Ground truth tests of the dual-wavelength radar detection of hail. Preprints, *16th Radar Meteor. Conf.*, Houston, Texas, Amer. Meteor. Soc., Boston, 41-42.

References

Eilts, M. D., 1987: Low altitude wind shear detection with Doppler radar. *J. Climate Appl. Meteor.*, **26**, 96-106.

Elder, F. C., 1957: Some persistent "ring" angles on high powered radar. *Proc. Sixth Weather Radar Conf.*, Amer. Meteor. Soc., Boston, 281-290.

Federal Aviation Administration, 1991: Introduction to ARSR-4, Chapter I: Technical System Overview, FAA USAF Radar Replacement Program (FARR), Federal Aviation Administration, Washington, 16 pp.

Fletcher, N. H., 1966: *The Physics of Rainclouds.* Cambridge, Cambridge University Press, 390 pp.

Foote, G. B., and P. S. DuToit, 1969: Terminal velocity of raindrops aloft. *J. Appl. Meteor.*, **8**, 249-253.

Frush, C. L., 1984: Using the sun as a calibration aid in multiple parameter meteorological radar. Preprints, *22nd Radar Meteor. Conf.*, Zurich, Amer. Meteor. Soc., Boston, 306-311.

Fujita, T. T., 1985: *The Downburst; Microburst and Macroburst.* Satellite and Mesometeorology Research Project Research Paper No. 210. Dept. of Geophysical Sci., Univ. of Chicago, 122 pp.

Fujita, T. T., and John McCarthy, 1990: Ch. 31a, The Application of Weather Radar to Aviation Meteorology. In: *Radar in Meteorology*, David Atlas, Editor. Amer. Meteor. Soc., Boston, 657-681.

References

Gossard, E. E., and R. G. Strauch, 1983: *Radar Observation of Clear Air and Clouds*. Elsevier, Amsterdam, Ch. 2.4.2-2.4.5.

Gunn, R., and G. D. Kinzer, 1949: The terminal velocities of fall for water droplets in stagnant air. *J. Meteor.*, 6, 243-248.

Gunn, K. L. S., and T. W. R. East, 1954: The microwave properties of precipitation particles. *Quart. J. Roy. Meteor. Soc.*, 80, 522-545.

Hamidi, Said, Ronald E. Rinehart, and John D. Tuttle, 1983: Test of a transverse-wind algorithm for NEXRAD in real-time. Preprints, *21st Conf. on Radar Meteor.*, Edmonton, Alberta, Canada, Amer. Meteor. Soc., Boston, 409-412.

Herzegh, P. H., and R. E. Carbone, 1984: The influence of antenna illumination function characteristics on differential reflectivity measurements. Preprints, *22nd Radar Meteor. Conf.*, Zurich, Amer. Meteor. Soc., Boston, 281-286.

Huschke, Ralph E., Ed., 1959: *Glossary of Meteorology*, Amer. Meteor. Soc., Boston, 638 pp.

JSPO, 1985: *Next Generation Weather Radar Algorithm Report*, NEXRAD Joint System Program Office, Norman, Oklahoma, ≈510 pp.

JSPO, 1986, *Next Generation Weather Radar Product Description Document*, NEXRAD Joint System Program Office, Norman, Oklahoma, ≈110 pp.

Levanon, Nadav, 1988: *Radar Principles*, John Wiley & Sons, New York, 308 pp.

Lhermitte, R. M., and D. Atlas, 1961: Precipitation motion by pulse Doppler radar. Proc. *Ninth Weather Radar Conf.*, Amer. Meteor. Soc., Boston, 218-223.

Marshall, J. S., and W. McK. Palmer, 1948: The distribution of raindrops with size. *J. Meteor.*, **5**, 165-166.

Matson, Richard, and Arlen W. Huggins, 1979: Field Observations of the kinematics of hailstorms. NCAR/CSD, Boulder, NCAR/TN-139+STR, 68pp.

McCormick, G. C. 1970: Reflectivity and attenuation observations of hail and radar bright band. Preprints, *14th Radar Meteor. Conf.*, Amer. Meteor. Soc., Boston, 19-24.

McCormick, G. C., and A. Hendry, 1970: The study of precipitation backscatter at 1.8 cm with a polarization diversity radar. Preprints, *14th Radar Meteor. Conf.*, Tucson, Arizona, Amer. Meteor. Soc., Boston, 225-230.

Mie, G., 1908: Beitrage zur Optik Truber Medien, speziell kolloidaler Metallosungen. [Contribution to the optics of suspended media, specifically colloidal metal suspensions]. *Ann. Phys.*, **25**, 377-445.

NEXRAD JSPO, 1986: Next Generation Weather Radar (NEXRAD) Product Description Document. NEXRAD Joint System Program Office, R400-PD-202.

References

Pike, J. M., and R. E. Rinehart, 1983: Calibration of a pressure sensor and a radar receiver using behavioral modeling. *J. Climate and Appl. Meteor.*, 22, 1462-1467.

Probert-Jones, J. R., 1962: The radar equation in meteorology. *Quart. J. Roy. Meteor. Soc.*, 88, 485-495.

Rinehart, R. E., 1978: On the use of ground targets for radar reflectivity factor calibration checks. *J. Appl. Meteor.*, 17, 1342-1350.

Rinehart, R. E., 1980: Mechanical speed and height calculators. Preprints, *19th Conf. on Radar Meteor.*, Miami Beach, Amer. Meteor. Soc., Boston, 523-528.

Rinehart, R. E., 1982: Out-of-level instruments: Errors in hydrometeor spectra and precipitation measurements. *J. Climate and Appl. Meteor.*, 22, 1404-1410.

Rinehart, R. E., 1979: Internal storm motions from a single non-Doppler weather radar. Ph.D. Dissertation, Colorado State U., Fort Collins, 262 pp.

Rinehart, R. E., 1984: Radar conferences and related trivia. Preprints, *22nd Conf. on Radar Meteor.*, Zurich, Amer. Meteor. Soc., Boston, 97-100.

Rinehart, R. E., 1989: The Rinehart Projection: A New Map Projection for Range-Aliased Data. Preprints, *24th Conf. on Radar Meteor.*, Tallahassee, Amer. Meteor. Soc., Boston, 697-700.

Rinehart, R. E., and P. J. Eccles, 1976: Use of a nodding dihedral target for antenna gain measurements.

Preprints, *17th Radar Meteor. Conf.*, Seattle, Amer. Meteor. Soc., Boston, 66-71.

Rinehart, R. E., and C. L. Frush, 1983: Comparison of antenna patterns obtained from near-field test measurements and ground target scans. Preprints, *21st Conf. on Radar Meteor.*, Edmonton, Amer. Meteor. Soc., Boston, 291-295.

Rinehart, R. E., and J. D. Tuttle, 1982: Antenna beam patterns and dual-wavelength processing. *J. Appl. Meteor.*, **21**, 1865-1880.

Sassen, Ken, and Liang Liao, 1996: Estimation of cloud content by W-band radar, *JAM*, **35**, 932-938.

Seliga, T. A., and V. N. Bringi, 1976: Potential use of radar differential reflectivity measurements at orthogonal polarizations for measuring precipitation. *J. Appl. Meteor.*, **15**, 69-76.

Skolnik, Merrill I., Editor, 1970: *Radar Handbook*, New York, McGraw-Hill Book Co.

Skolnik, Merrill I, 1980: *Introduction to Radar Systems*, New York, McGraw-Hill Book Co., 581 pp.

Taylor, Albion D., 1981: The NOAA solar ephemeris program, NOAA Tech. Memo. ERL ARL-94, Environ. Resch. Lab., Nat. Oceanic and Atmos. Admin., U.S. Dept. of Commerce, Boulder, 28 pp.

Taylor, John R., 1982: *An Introduction to Error Analysis*. University Science Books, Mill Valley, CA, 270 pps.

References

Trammell, Archie, 1989: *Airborne Weather Radar Pilot's Operating Guide.* AJT, Inc., Trinidad, TX. 30 pps.

U. S. Naval Observatory, ~1995: *MICA for DOS, 1990-1999, Version 1.0, User's Guide.* Astronomical Applications Dept., U.S. Naval Observatory, 3450 Massachusetts Ave., N.W., Washington, D.C. 20392-5420.

Wexler, R., and D. Atlas, 1963: Radar reflectivity and attenuation of rain. *J. Appl. Meteor.,* **2,** 276-280.

Whiton, R. C., P. L. Smith, and A. C. Harbuck, 1976: Calibration of weather radar systems using the sun as a radio source. Preprints, *17th Conf. on Radar Meteor.,* Seattle, Amer. Meteor. Soc., Boston, 60-67.

Index

A

B

Index

Index

Index

F

G

H

I

Index

K

L

M

Index

N

O

P

Q

R

Index

S

Index

W

Z